Beyond Baby M

Contemporary Issues in Biomedicine, Ethics, and Society

Beyond Baby M: *Ethical Issues in New Reproductive Techniques*, edited by *Dianne M. Bartels, Reinhard Priester, Dorothy E. Vawter,* and *Arthur L. Caplan, 1989*

Clinical Ethics: *Theory and Practice*, edited by *Barry Hoffmaster, Benjamin Freedman,* and *Gwen Fraser, 1989*

Reproductive Laws for the 1990s, edited by *Sherrill Cohen* and *Nadine Taub, 1989*

What Is a Person, edited by *Michael F. Goodman, 1989*

Advocacy in Health Care, edited by *Joan M. Marks, 1986*

Which Babies Shall Live?, edited by *Thomas H. Murray* and *Arthur L. Caplan, 1985*

Feeling Good and Doing Better, edited by *Thomas H. Murray, Willard Gaylin,* and *Ruth Macklin, 1984*

Ethics and Animals, edited by *Harlan B. Miller* and *William H. Williams, 1983*

Profits and Professions, edited by *Wade L. Robison, Michael S. Pritchard,* and *Joseph Ellin, 1983*

Visions of Women, edited by *Linda A. Bell, 1983*

Medical Genetics Casebook, by *Colleen Clements, 1982*

Who Decides?, edited by *Nora K. Bell, 1982*

The Custom-Made Child?, edited by *Helen B. Holmes, Betty B. Hoskins,* and *Michael Gross, 1980*

Medical Responsibility, edited by *Wade L. Robison* and *Michael S. Pritchard, 1979*

Contemporary Issues in Biomedical Ethics, edited by *John W. Davis, Barry Hoffmaster,* and *Sarah Shorten, 1979*

Beyond Baby M

Ethical Issues in New Reproductive Techniques

Edited by

Dianne M. Bartels,
Reinhard Priester, Dorothy E. Vawter,
and Arthur L. Caplan

Center for Biomedical Ethics, University of Minnesota,
Minneapolis, Minnesota

Springer Science+Business Media, LLC

Library of Congress Cataloging in Publication Data
Main entry under title:

Beyond Baby M: ethical issues in new reproductive techniques/edited by Dianne M.
   Bartels...[et al.]
288 p.   6 x 9 in.—(Contemporary Issues in biomedicine, ethics, and society)
    ISBN 978-1-4612-8853-4   ISBN 978-1-4612-4510-0 (eBook)
    DOI 10.1007/978-1-4612-4510-0

    1. Human reproductive technology—Moral and ethical aspects.
I. Bartels, Dianne M.   II. Series
RG133.5.B49   1989
176—dc20   89-49394
                                                        CIP

Permission to reprint *Surrogacy: A Question of Values* was obtained from *Conscience*
Permission to adapt *Recreating Motherhood: Ideology and Technology in a Patriarchical Society* (1989) was obtained from W. W. Norton.

© 1990 Springer Science+Business Media New York
Originally published by Humana Press Inc. in 1990
Softcover reprint of the hardcover 1st edition 1990

All rights reserved

No part of this book may be reproduced, stored in a retrieval system, or transmitted in any form or by any means, electronic, mechanical, photocopying, microfilming, recording, or otherwise without written permission from the Publisher.

Contents

vii Contributors
viii Acknowledgments

1 Introduction
   Arthur L. Caplan

Society and Reproductive Issues

9 Recreating Motherhood:
   Ideology and Technology in American Society
   Barbara Katz Rothman
29 Science, Conscience, and Public Policy:
   Historical Reflections on Controversial Reproductive Issues
   John Eyler
45 Current Religious Perspectives
   on the New Reproductive Techniques
   Baruch Brody
65 Essential Ethical Considerations for Public Policy
   on Assisted Reproduction
   Carol Tauer

Treatment of Infertility and Assisted Reproduction

89 Medical Techniques for Assisted Reproduction
   George Tagatz
111 Infertility and the Role of the Federal Government
   Gary Ellis
131 Sexuality and Assisted Reproduction:
   An Uneasy Embrace
   Kathy J. Harowski

149 Arguing with Success:
Is In Vitro Fertilization Research or Therapy?
Arthur L. Caplan

Surrogate Motherhood

173 Surrogacy Arrangements: *An Overview*
Dianne M. Bartels
183 The Case of Baby M
Janna Merrick
201 Surrogate Motherhood Agreements:
The Risks to Innocent Human Life
James Bopp, Jr.
221 The Vatican *Instruction* and Surrogate Motherhood
Richard Berquist
235 Surrogacy: *A Question of Values*
Barbara Katz Rothman
243 Surrogacy and the Family:
Social and Value Considerations
Adrienne Asch

261 Appendices
Baby M Contract, 261
State Legislative Examples:
Michigan, 269
Nevada, 272
New York, 273
Vatican Statement, 277

285 Index

Contributors

ADRIENNE ASCH • *New Jersey Bioethics Commission, Trenton, NJ*
DIANNE M. BARTELS • *Center for Biomedical Ethics, University of Minnesota, Minneapolis, MN*
RICHARD BERQUIST • *Department of Philosophy, College of St. Thomas, St. Paul, MN*
JAMES BOPP, JR. • *Tudor House, Terre Haute, IN*
BARUCH BRODY • *Center for Ethics, Medicine and Public Issues, Baylor College of Medicine, Houston, TX*
ARTHUR L. CAPLAN • *Center for Biomedical Ethics, University of Minnesota, Minneapolis, MN*
GARY B. ELLIS • *Division of Health Promotion and Disease Prevention, Institute of Medicine of the National Academy of Sciences, Washington, DC*
JOHN EYLER • *History of Medicine, University of Minnesota, Minneapolis, MN*
KATHY J. HAROWSKI • *Program in Human Sexuality, University of Minnesota, Minneapolis, MN*
JANNA MERRICK • *Department of Political Science, St. Cloud State University, St. Cloud, MN*
BARBARA KATZ ROTHMAN • *Department of Sociology, Baruch College, The City University of New York, New York, NY*
GEORGE TAGATZ • *Department of Obstetrics/Gynecology, University of Minnesota, Minneapolis, MN*
CAROL TAUER • *Department of Philosophy, College of St. Catherine, St. Paul, MN*

Acknowledgments

The editors give special acknowledgment and thanks to Patty Vogt at the Center for Biomedical Ethics, whose ability to translate our scrawlings and to organize the disorganized made this production possible. We would also like to thank Janet Callen and Julie Lifto, who provided major assistance in organizing the conference that inspired these papers. Candy Holmbo, Paul Welvang, Mara Nervick, Bill Sucha, and Barbara Higgins have assisted and encouraged us throughout this project. We are grateful for such a supportive team of people in the Center for Biomedical Ethics.

Financial and technical support was generously provided by Bart Galle and the Continuing Medical Education staff; David M. Brown, Dean of the Medical School; and Geoffrey Kaufmann, University of Minnesota Hospital and Clinic. Naomi Scheman, Department of Philosophy and Women's Studies, was most helpful in editing selected texts in the area of feminist ethics. We wish to thank Karen Grandstand Gervais for developing the Index. We also appreciate Noel Keane's willingness to share the Baby M contract, and *Conscience*'s editorial staff's permission to reprint "Surrogacy: A Question of Values."

Finally, and most importantly, we thank the authors for their patience and participation in the editing process. We hope this final product is a worthy outcome for the extensive work that went into it.

Dianne M. Bartels
Reinhard Priester
Dorothy E. Vawter
Arthur L. Caplan

Introduction

Arthur L. Caplan

It is commonly said, especially when the subject is assisted reproduction, that medical technology has outstripped our morality. Yet, as the essays in this volume make clear, that is not an accurate assessment of the situation.

Medical technology has not overwhelmed our morality. It would be more accurate to say that our society has not yet achieved consensus about the complex ethical issues that arise when medicine tries to assist those who seek its services in order to reproduce. Nevertheless, there is no shortage of ethical opinion about what we ought to do with respect to the use of surrogate mothers, in vitro fertilization, embryo transfer, artificial insemination, or fertility drugs.

Nor is it entirely accurate to describe assisted reproduction as technology. The term "technology" carries with it connotations of machines buzzing and technicians scurrying about trying to control a vast array of equipment. Yet, most of the methods used to assist reproduction that are discussed in this volume do not involve exotic technologies or complicated hardware. It is technique, more than technology, that dominates the field of assisted reproduction.

Efforts to help the infertile by means of the manipulation of human reproductive materials and organs date back at least to Biblical times. Human beings have engaged in all manner of sexual practices and manipulations in attempts to achieve reproduction when nature has balked at allowing life to begin.

Part of the reason it seems as if ethics is desperately playing a game of catch-up with contemporary treatments for infertility and assisting reproduction is that our society has, for a variety of reasons, chosen to avoid attempts at reaching moral consensus. In a pluralistic society, such as that of the United States, it is not always possible to achieve ethical consensus when strong differences of opinion exist about fundamental moral questions; and reproduction and sexuality are areas in which fundamental differences of moral opinion certainly exist.

Our leaders and key government bodies have consciously steered away from direct attempts to grapple with the ethical challenges raised by advances in assisting reproduction. For more than a decade, the federal government has been unwilling to take a public position on the use of eggs and embryos in research or therapy. It is only in the past few years that state and federal officials have become willing even to hold hearings or commission studies on such topics as surrogate mothering or the efficacy of in vitro fertilization.

The avoidance of a direct confrontation with the ethical challenges of assisted reproduction may represent a collective failure of leadership on the part of the American government. However, it may also represent a recognition, even if it is a tacit recognition, that, when no clear-cut consensus exists, it may be simply wise to avoid efforts at legislation and regulation, in order to allow debate and discussion to proceed. It is hard to proceed in as sensitive an area as reproduction without clear-cut rules and standards. It may be even harder to impose clear-cut rules and standards when sincere disagreement exists as to the best course of action and public policy with respect to reproduction—an aspect of life into which government, the judiciary, and bureaucracy have been loathe to tread.

Surrogacy, in vitro fertilization, gamete interfallopian transfer, and embryo transfer are products of medical and scientific advances in understanding some of the ways in which human reproduction works. However, they are also, as many of the contributors to this volume note, products of a distinctive set of social and economic changes in American society. Decisions to delay childbearing, a rise in infertility associated with sexually transmitted diseases, the rise in divorce rates, the entry of a large number of women into the workplace, the changes that have reshaped the nature of the nuclear family, and the general availability of contraception and abortion have all set the stage for a demand for assistance in reproducing.

Although the questions raised by cases, such as that of Baby M, or the birth of children, such as Louise Brown, by means of in vitro fertilization are important, they represent the tip of the iceberg in terms of the moral issues that American society will have to address in the years to come. We have not yet decided whether infertility is to be treated as a disease, a disability, or simply an inconvenience. We still have no consensus as to whether it is necessary to have medical supervision over the use of such techniques as artificial insemination or surrogacy. There has been little attempt to grapple with the question of the legal standing of embryos, fetuses, or children created by means of unusual modes of conception.

Most importantly, there has been no serious effort to grapple with what is perhaps the toughest ethical question of all—who should have the ability, or even the right, to avail themselves of techniques to assist reproduction?

Should assisted reproduction be restricted to those who can satisfy a physician that they are incapable of creating life by means of heterosexual intercourse within marriage, or should the techniques be available to anyone who might wish to use them, including single men and women, gays, those who might find pregnancy an inconvenience or a psychological burden, or those who face a risk to their health if they become pregnant, as appeared to be the motivation in the Baby M case?

The most disturbing ethical question regarding eligibility for using the techniques of assisted reproduction is whether it should be available to those who want to use the techniques to try to improve the health, the abilities, or the physical traits of their offspring. With the emergence of genetic engineering, it will not be long before the techniques of assisted reproduction can be combined with those of genetic engineering to create the possibility of not only assisting reproduction, but also manipulating the traits and properties of the children created by these techniques.

It is one thing to debate the morality of using medical knowledge to assist those who cannot have children because they are infertile. It can be fascinating to follow the ins and outs of dilemmatic cases, such as Baby M. Yet, it is a very different, but much more significant question to debate the morality of using assisted reproduction in combination with genetic knowledge to try to improve the human species.

The first rumblings of the morality of eugenics are on the horizon. Some parents have sought to treat their children with growth hormone, which can now be created in large quantities by means of genetic engineering. They want to use the hormone, not because their children have congenital impairments or diseases that adversely affect

Introduction

their potential for growth, but because they admire height or believe that being tall is advantageous in economic and social terms in our society. These are the first inklings of the social and public policy choices that will confront our society as our ability to design our descendents expands in the years to come.

The cases and controversies described in the essays in this book are, obviously, interesting in their own right. Who should get custody of a child when a surrogate changes her mind? Should we allow commercial surrogacy, or ban it as incompatible with central social values concerning commerce and the human body? Should we allow scientists to conduct experiments on human embryos, created as a part of standard in vitro fertilization techniques, that otherwise will simply be destroyed?

The views and arguments presented in this volume should be considered as much for their long-term implications as for their value in locating answers to the questions that lie just around the corner in the realm of reproduction. We need to begin talking now about what sorts of families we believe our society should encourage and foster, what sort of priority we ought to give to making resources available to all citizens to choose assisted reproduction if they desire it, and what sorts of reasons constitute ethical reasons for using assisted-reproduction techniques as a way to create life.

Government has appeared to be avoiding the questions raised by advances at the frontiers of assisting reproduction for a long time. Yet, it might be more accurate, and certainly more generous, to say that our society understands the enormous significance of the ethics of reproduction and wants to proceed with caution in order to achieve consensus. Where matters of morality and medicine are concerned, society is best served not by policies based on fear, ignorance, prejudice, or raw emotion, but by the emergence of moral consensus. Some of the building blocks for achieving consensus can be found in the pages of this book.

Society
and
Reproductive Issues

Recreating Motherhood

Ideology and Technology in American Society*

Barbara Katz Rothman

The science and technology that a society develops reflect the values of its culture. This is as true in what we call "reproductive technology" as it is in the "space industry." Guides to the culture are most clearly found in the language used: as has often been noted, it is significant that we use words like "conquering" and "frontier," and even the very word "space" as if it were empty without us. So it is with "reproduction," a word that implies that making babies is a form of production—raw material transformed by work into a product.[†]

In the past few years, as I have participated in the conferences that have been springing up on "New Reproductive Technology," reflecting the growing American concern and fascination with the topic, I have paid close attention to the language being used. I have heard about "state-of-the-art" babies and about "harvesting" eggs from women's bodies. I have also heard women's bodies themselves

---

*This article is drawn from the author's book *Recreating Motherhood: Ideology and Technology on a Patriarchal Society*, W. W. Norton, 1989.
†My appreciation to Ruth Hubbard for pointing out the inappropriateness of the term "reproduction." As she says it: we do not reproduce ourselves; two of us come together and make a third one of us.

called "maternal environments." I have heard abortions of wanted and loved pregnancies called "therapeutic" because there was no therapy available for the condition that the fetus had. I have heard miscarriages of desperately wanted pregnancies called "successes" because a pregnancy was "achieved" with in vitro fertilization.

I do not think that this language we are using reflects the experiences of women as mothers. It more directly reflects the concerns and interests of the reproductive technologies themselves, and at a deeper level, the perspective of men in a patriarchal society.

The term "patriarchy" is often used loosely as a synonym for "sexism" or to refer to any system in which men rule. The term technically means "rule of fathers," but in its current practical usage more often refers to any system of male superiority/female inferiority. However, male dominance and patriarchal rule are not quite the same thing, and when we look at the values surrounding procreation, the differences are important.

Patriarchal kinship is the core of what is meant by patriarchy: the idea that paternity is the defining, central social relationship. In a patriarchal kinship system, children are reckoned as being born to men, of their line, out of the bodies of women. Women, in such a system, bear the children of men: Mrs. John Smith bears John Smith, Jr.

The essential concept here is the "seed"—the part of men that grows into children. Early scientists in Western society were so deeply committed to the patriarchal concept that it influenced what they saw. One of the first uses of the microscope was to look at semen to see the little person, the homunculus, curled up inside the sperm. In 1987, the director of a California sperm bank distributed T-shirts with a drawing of sperm swimming on a blue background and the words "Future People."

Out of the patriarchal focus on the seed as the source of being—on the *male* production of children from sperm—has grown our current, usually more sophisticated thinking about procreation. Modern science has had to confront the *egg* as seed also. The doctor who has spent time "harvesting" eggs from women's bodies for in vitro fertilization fully understands the significance of women's seeds. Whereas the old patriarchal system had a clear place for women, as the nurturers of men's seeds or the soil in which seeds grow, that is no longer acceptable. By recognizing the significance of women's genetic contribution to the next generation, modern Western patriarchy extended to women some of the privileges of patriarchy. That is, when the significance of women's seed is acknowledged in her relationship with her children, women too come to have *paternity* rights in their children. In this modified system based on the older ideology of patriarchy, women too can be seen to own their children, just as men do. This relationship between women and their children is not based on motherhood *per se*, not on the unique nurturance, the long months of pregnancy, the intimate connections with the baby as it grows and moves inside her body, passes through her genitals, and sucks at her breasts. Instead, women are said to own their babies and have "rights" to them, just as men do: based on their seed.

The new procreative technology being developed is based on this focus on the seed. The seed—the genetic material—is the one absolutely irreplaceable part of procreation as our science now approaches it. The technology we are developing continues to substitute for one after another of the nurturing tasks, but makes no substitute for the seed. Breasts became unnecessary quite some time ago, as artificial formula substituted for human milk. The act of giving birth became increasingly unnecessary as doctors worked on surgical removal of babies, to the point where more than one out of five American babies are born by Caesarean section. The nurturance of late pregnancy became unnecessary as neonatal intensive care units developed the skill to maintain younger and younger, smaller premature babies or "extrauterine fetuses" in incubation, and the nurturing environment of the fallopian tubes became replaceable as the nurturance of the glass dish, the in vitro environment, was developed.

None of these techniques of artificial nurturance work as well as the natural mothering experience, though as one or another becomes "faddish," there is a tendency for doctors to proclaim its superiority over the natural mother. Most importantly, the mother herself becomes reduced to the background, the environment, the site in which genetic material develops. Hence, we begin to think about the relative advantages and disadvantages of the various sites for genetic development to occur: human vs artificial or animal wombs.

Nevertheless, a pregnant woman is not a *place*. She is not a walking incubator or a site. She is a person, engaged on every level in the process of procreation. A growing, developing embryo/fetus/baby is not "housed" or "stored" in the process of its development; it too is increasingly engaged on many levels: physical, psychological, social, and emotional. Between the moment of zygotic zero, when sperm and egg fuse and genetics are set, and the time of birth, there have been months of development. Even at birth, babies have a history—months of experience that shape who they are and what they bring to their new life.

The kind of technology that we are developing encourages us to dismiss the importance of nurturance over "seed" and to fail to see the significance of gestation as an experience for the potential child as well as for its mother.

In certain ways, we act as if the child first springs into being at birth. Consider the differences in meaning between the archaic "with child" and the more contemporary "expecting." It is this kind of thinking that encourages us to develop both technology and marketing that focus on getting seeds to grow and harvesting state-of-the-art babies. The reduction of mothering to a site for fetal development is what makes "surrogacy" a thinkable thought.

In contrast to this view, I am going to focus now on a view of pregnancy not as a place or even a time, but as a *relationship*. Human beings are not chromosomes grown up—not sperms and eggs writ large. If we are "products," we are the products of our experiences and our relationships. Children do not enter the world from outside the world; they do not come from Mars or out of a black box. By the time they are born, they have been here, in this world, for nine months: not as children, not as people, but as part of their mothers. Women giving birth do not experience the "arrival" of babies. Instead, they feel them leave. A baby is born already in a relationship—a physical, social, and emotional relationship with the woman in whose body it was nurtured.

The Physical Relationship

Medicine and science have relatively recently recognized the significance of the physical environment provided in pregnancy. It was not until the late 1800s that physicians began to provide prenatal care. Before that time, routine medical supervision was not considered necessary.[1] How the baby came out, that it could be hurt on the way out—doctors were interested in this while they still believed that, as long as the baby was in the womb, it was beyond the influence of the environment, and consequently, beyond the influence of doctors and medical care.

It took the thalidomide tragedy to destroy that idea completely. The birth of hundreds of babies with serious, obvious, physical impairments, directly traceable to a drug mothers took (a drug doctors prescribed for nausea in pregnancy) changed forever the medical view of the uterus and the meaning of pregnancy. The placenta, once thought of as an impenetrable barrier, came to be seen as a "bloody sieve."

However, the more things change, the more they stay the same. Doctors could have used this new information to come to the same conclusion that I have come to: the fetus is part of its mother's body, no safer than any of her organs from damage done to her. Like every other part of her body, the fetus has special susceptibilities and special resiliences. That is, poisons that damage the kidney may do no direct damage to the heart. Things that do direct damage to the heart may not directly harm the kidney. Yet the whole is connected: ultimately, poisoning someone's kidney or heart is poisoning the person. Some organs we can live without—one kidney is adequate. Some we cannot—we have only one heart. The fetus we can live without, so it is possible to destroy the fetus and not kill the mother. Also, for most of the pregnacy, the fetus is enough like an organ that it is impossible to damage the mother seriously and not damage the fetus.

Nevertheless, the medical profession did not come to the conclusion that fetuses are fundamentally a part of their mothers' bodies. They stayed instead with the essentially patriarchal view that fetuses are seeds growing up—entirely separate beings "planted" in the mothers. Now, however, instead of seeing the mother as a safe haven, they began to see that there was, as the old spiritual says, "no hiding place down here." Previously, the uterus was thought of as a protected nest, but at this point in time, it came to be viewed as unsafe and inadequately protected.

The fetus was discovered to be vulnerable to harm, as we all are, and the all-powerful, protective mother had failed.

Before this fall from grace, when the mother's body was believed to do the work of protection, the mother's mind was of little importance. Once the mother's body could no longer be trusted to protect the fetus, and the nest was shown to be open to danger, then the mother's job became guarding the nest. The mind, as distinct from the body, became important. Women had to know consciously what the fetus needed and what would harm the fetus. Women had to be taught how to nourish and how to protect their fetuses; also, women had to be willing to learn.

The notion that there are dangers that pregnant women must avoid is, of course, not at all new. All cultures seem to have pregnancy taboos. What Western medicine did was to make the shift from dismissing all taboos as "old wives' tales" to instituting its own taboos.

The more that doctors learned about the unique vulnerabilities of the fetus, the more importance they placed on "compliance" or, in other words, on the willingness of a patient to "follow doctor's orders." The focus on compliance actually started before thalidomide, when doctors feared toxemia. Toxemia is a disease of late pregnancy, the causes of which are still not clear. The 1980 edition of *William's Obstetrics* notes that everyone from "allergist to zoologist" has proposed a theory.[2] The symptoms are a dramatic rise in blood pressure, the accumulation of enormous amounts of water, so that the woman swells up, and the spilling of protein into urine. Unchecked, toxemia can cause convulsions, which can either directly kill the fetus or bring about a premature birth. The mother too can die. Prenatal care is organized around screening for toxemia: blood pressure along with weight gain checks, and urinalysis are all that comprise a standard monthly visit. Since a symptom of toxemia is the sudden weight gain from all the accumulated water, doctors began to check weight very carefully. Somehow, the idea of weight gain as a *symptom* slipped over into weight gain as a *cause*, and doctors began to put great emphasis on women keeping their weight down during pregnancy. By the 1950s and 60s, it had gotten so out of hand that medical authorities were recommending that overweight women lose weight while pregnant. The July 15, 1962 issue of the *American Journal of Obstetrics and Gynecology* had four full-page ads for drugs to be used to control weight in pregnancy—drugs including phenobarbitol and amphetamines to suppress appetite.[3]

There were important social and psychological effects of this medical emphasis on weight. Even now that the idea of so severely restricting weight to prevent toxemia has been totally discredited, we are left with its residue of distrust.

Doctors had set up a nearly unachievable goal. It was very difficult for women with adequate access to food to keep pregnancy weight gain down to no more than the prescribed 24 pounds. It was virtually impossible for women who were a few pounds overweight to keep their weight gain down to even less, much less to lose weight while pregnant. So all through the course of pregnancy, women under medical care—virtually all middle-class women—had the experience of a monthly evaluation of their competence as mothers, and the majority failed. Doctors, believing that women were putting their health and their babies at risk by overeating, began to see women as foolishly self-indulgent.

What happened was that, by setting up an unattainable—and as it turns out entirely inappropriate and ultimately dangerous—goal, both mothers and doctors learned not to trust mothers. Try as they might, women could not put what they were told—what they genuinely believed, was in the best interests of their babies into action, against the messages they were getting from their bodies. They learned not to trust their bodies. The doctors, for their part, learned that women could not be relied on to follow their advice as to what their babies needed.

The thalidomide disaster occurred in 1961, while this battle of the bulge was still being fought. The cultural lesson that was learned (the lesson that passed into popular knowledge) was not what now seems most obvious: that doctors can make terrible, terrible mistakes. In the United States, there was one woman, Frances Kelsey, in the FDA who was concerned about chick embryo studies that showed that thalidomide interfered with limb development. By preventing its sale here, she not only saved countless babies from loss of limbs, but she also saved the American medical profession from a serious loss of face. Had she let thalidomide through, it is hard to believe that American doctors would have been any more reluctant than their European counterparts to prescribe it, or more reluctant to prescribe thalidomide than DES, Bendectin, or any number of other drugs since proven dangerous—including the heavily advertised diet drugs.

However, the fact remains that, in the US, women did not get thalidomide from their doctors. Hence, the message that came through was not that doctors can be hazardous to pregnancy, but that *mothers* can be. Women can take things that will hurt their babies. Not only are mothers not protective—mothers are a potential source of harm. Babies need protection, not *by* their mothers, but *from* their mothers.

The legacy with which we are left is a sense of mothers and their fetuses being at odds with each other, and doctors being the mediators. Even while our cities cut funds for services to pregnant women, they run ad campaigns to show how dangerous women are to their fetuses, stressing individual failure to seek prenatal care and lack of good nutrition in pregnancy as prime causes of infant mortality. Going even further, various states have begun to contemplate, and on occasion pursue, legal action against pregnant women who do not obey their doctors' orders. It seems that we forget or just ignore that, over all the years of evolution, it has been mothers who have most consistently protected and sacrificed for their fetuses and babies. We forget or ignore that just as women are not the natural enemies of the fetus, doctors are not their natural protectors. It was, after all, the obstetricians who fought against "natural childbirth," the movement, beginning in the 1950s, of mothers who wanted undrugged births and undrugged babies: their doctors wanted compliant patients and were quite willing to put fetuses at risk to ensure the compliance of their mothers.

Doctors are not all-knowing. The people who have brought us routine X-rays in pregnancy, diuretics, DES, and bendectin, do not always know what they are doing. Furthermore, people do not always take the best possible care of themselves. They do not always eat wisely, exercise appropriately, and avoid hazardous substances and places. This is probably no less, but certainly no more, true in pregnancy than at any other time.

The Social Relationship

Mothers and fetuses are not just chemically connected. The placenta is the point at which their fluids meet, but there are other points of meeting: heads rest on bladders in pregnancy, and feet dance against ribs. The mother holds her fetus within her.

The relationship between a mother and her fetus is not just a chemical relationship—an exchange of nutrients, hazards, and wastes—or a purely "mechanical relationship." Contemporary medical language would have us draw on all kinds of mechanical metaphors for the physical relationship between the mother and her fetus, but neither is a machine. It is not a question of so many centimeters of head circumference fitting into so many centimeters of pelvis.

The mother, as a social being, is responding socially to the experience of carrying her baby. Her reactions are not mechanical; they are social. I am not referring here to women reading Latin poetry to their fetuses in the hopes that they will enter Harvard on scholarship. I am saying that *any* mother is engaged in a social interaction with her fetus as the pregnancy progresses. The pregnancy is a social as well as a physical relationship. Women are not "flowerpots" in which babies are planted,[4] but social beings, giving social meanings to their experiences. When her baby uses her bladder as a trampoline, a woman responds. She responds not only by making another trip to the bathroom. She responds socially, with annoyance, amusement, irritation, anger, sometimes even pleasure at the apparent liveliness of the baby, and most often by the end of pregnancy, with a longing to end this phase of the relationship, but respond she does, not only to the physical experience, but also to the social and emotional overlays of meaning given to that experience.

Negotiating sleep in the last weeks of pregnancy is an even clearer example. It is not an easy thing to fall asleep with a six-pound jumping bean in your belly; but the fetus is not simply a mechanical irritation. Mothers need to find ways of soothing fetuses to sleep when they want to sleep. With a great deal of thought, or with no particular conscious effort at all, women learn how to lie in bed and what to do, to ease the fetus into quiet—or when to give it up and get out of bed for a while.

The fetus, for its part, is not yet a social being; these interactions with its mother are its first social experiences. In acting as if the baby "arrived" from outside—"entered" the world—we are acting as though children start as separate people and arrive in our lives as babies, but there is a continuum from the single cell to the newborn child to the youngster. The fetus/baby/child's actions affect others, who respond socially. In the course of these interactions, the child eventually becomes a social being as well—someone with a sense of self. Women's experience of this growth from a cell to a person is continuous, and men's discontinuous—in goes a seed, and out comes a baby.[5] As a patriarchal culture, by making men's reality our ideology, we deny the continuity that women experience, and in denying it, we violate the continuity. By acting as if it were not there, we destroy what was there.

Again, let me take the relatively simple example of sleep. In the earlier part of pregnancy, the embryo/fetus has no sleep/wake experience that we could recognize. It cannot, until it has developed sufficient brain capacity. However, by 33 weeks—about two months before they are born—fetuses have developed the cycle of two states of sleep that adults have, rapid eye movement (REM) sleep in which we dream, and deep (non-REM) sleep. As the fetus takes up more and more space in its mother's body and it becomes more capable of responding to its environment, mother and fetus have to come to some kind of shared cycle. Although babies can, we all know, sleep through anything, presumably the mother jumping up to a loud alarm clock, taking a shower, listening to music, banging pots around in the kitchen, and taking a subway ride to work are bound to have some effect on the fetus. Sounds do penetrate, and movement of course is felt. On the other hand, when she lies down at night, the mother does what she can to get the baby to sleep. Lying on her back presses the baby against her spine: a lumpy mattress against which the fetus may protest. She may learn—she may be taught by her fetus—to lie on her side. Her slowed, even breathing and her relaxation ease her fetus to sleep. They accommodate each other.

This accommodation goes on through the night. No one "sleeps through the night." We all have periods of relative wakefulness and periods of deeper sleep. Sleep researchers have shown us that there are regular cycles to deep sleep, REM, and wakefulness. In pregnancy, mother and fetus have to share the same rhythms. When babies are born into their mothers' own beds, there is a continuity in this sleep/wake cycle. Right at the time of birth, both babies and mothers, if undrugged of course, are wide awake. In babies, this initial period of quiet alertness is a very special time and part of what all the fuss about immediate "bonding" is about. Shortly after birth, both mother and baby are generally sleepy.

When, from birth, babies sleep by their mothers' sides, as is the case throughout most of the world, the transition from inside to outside is part of a continuum of change. The sleep/wake cycle of mother and baby can stay "in sync." In American society, it is customary for babies to be born in hospitals and put immediately onto hospital schedules. For a few days, mother and baby are kept separate, particularly at night. When they are "brought home," as if for the first time, babies are often put into separate rooms, almost always into separate beds even if in the same room. Any synchronization of sleep cycles that was there is gone—destroyed because it was discounted. The mother finds herself awakened out of deep sleep at seemingly random times during the night. New parenthood in America is experienced as above all else an exercise in sleep deprivation. That is not a biological fact of life: that is a cultural creation.

If we were to recognize the continuity—the continuing connection between a mother and her fetus/baby—we would not destroy their intimate rhythms. We would not treat the baby as if "delivered" from outside, and bring that baby home from the hospital as if it came from the hospital to start with. However, American mothers are specifically told not to take the baby to bed. We try to avoid intimacy (any hint of sexual intimacy) in the most profoundly intimate and essentially sexual of experiences. We ignore the fact that the baby has been sleeping in its mother's bed, and in its mother's body, all along. We act in every way as if the baby were foreign, and we make of it a foreigner—an alien, a "little stranger." Yet it is not alien; it is part of its mother—a newly separated part—coming into its own separate existence.

The fetus enters the world not only as part of its mother's body, but also as part of the life the mother lives, part of the rhythms of her day and of her household, as well as of her body. Again, starting with the simple example of sleep/wake cycles, throughout the day as well as the night, pregnant women have periods of activity and periods of rest. Fetuses are *there*, sharing these experiences. When they "arrive" on the outside, they come with this shared background. One of the clearest examples is the dinnertime experience of so many new parents. Before the baby was born, the time right before dinner was consistently, for many women, one of the most active times of the day. Most women seem to retain responsibility for the kitchen and cooking, whether they have outside jobs or not. Getting dinner ready, whether it is following a commute from work or while chasing a toddler around the house, or both, tends to be a busy, noisy time for mothers in many households. Pots bang, water runs, and the mother is calling out to children or other members of the household. Lo and behold, after the baby is born, into this general bedlam, as sure as God made little green apples, comes a wide-awake baby. If parents try to fight it, put the baby into a quiet room, dark and alone, and hope it will go to sleep, they sometimes succeed, but often end up with a crying baby on their hands. People seem to have more success with moving the baby right into the middle of the action *where it is used to being*, either in a baby carrier strapped to a parent, or propped on the kitchen table.

The baby outside is not an entirely different creature than the baby inside. We may deny the continuity, but we live with its effects. Patterns learned inside are expressed outside—not only sleep/wake cycles, but even habits. Some babies learn to suck their thumbs *in utero* and continue doing so; some do not and may never learn. What the baby was inside is part of what the baby is outside. Its capacities for social interaction begin to develop in the months before birth.

By 32 weeks, the fetus' hearing functions much as it will after birth. The fetus hears and responds to speech, learning the rhythm of human speech long before the words. Babies show movement (almost a dance) to the sounds of speech, and newborns can recognize their mothers' voices,—something that mothers have often felt to be true, but knew could not be true. Now we have experimental evidence supporting this knowledge, so now we can indeed know what we know. A baby of less than three or four days old prefers a recording of its mother's voice reading a story to another woman's voice reading the same story. In an experiment, babies were given "pacifiers" to suck. Depending on their rate of sucking, they heard one of the two recordings. Within 20 minutes, babies learned what rate of sucking provided their mothers' voices.[6]

These late-pregnancy experiences of the fetus shape the baby, just as the experiences of a baby shape the child it becomes. Some of these are near-universal experiences

—they shape the near-universal experience of babycare—the singing, rocking, swinging, swaddling, and holding close that show up in all cultures in various forms.

Babies are soothed by singing; singing is probably most like the way fetuses hear their mothers' and other voices—strong on rhythm, weak on distinct words. Babies are soothed by being held firmly; the uterus was a tight swaddle, and the contractions of pregnancy provide frequent "hugs." Babies are soothed by being held close to the heart. By now it is common knowledge that babies calm down to the sounds of the adult—72 beat per minute heart rate: this has been turned to profit-making by selling teddy bears that play the recorded heartbeat. Babies are soothed by rocking: the rocking chairs, cradles, and hammocks used all over the world duplicate the rock of the pelvis when the mother walked in pregnancy. Babies seem to *need* to be rocked or walked: they are coming out of an environment in which they had been rocked for hours each day.

When the new reproductive technologists talk about creating artificial wombs, they are talking about creating a very different environment for a fetus. What would a baby be like that had grown without being held close, without an adult heartbeat surrounding it, without the sound of speech, and without being rocked? On the other hand, would the artificial womb come provided with a mechanical mother: some rocking, a recorded heartbeat, recorded voices (saying what?), and artificial contractions of the artificial womb? Grown outside of a woman—outside of the human community—we could indeed create an alien baby—a little stranger, the living reification of our ideology.

It is with mothers, of course, that fetuses have the most direct relationships, and it is on mothers that fetuses have the most direct effects: when they move, the mothers feel them. However, it is not only to mothers that fetuses relate. The movement of the fetus within is also felt by people in physically intimate contact with the mother, especially her lover and other children. The closer—literally, physically closer—other people are to the mother, the closer they are, physically and socially, to the fetus. When they hold the woman close, they too feel the fetus within. It reaches out to them, and that too is a social experience. It is also an interaction. Surely it is a very different experience to be the lover to a woman whose fetus lies with the limbs facing outward, kicking, prodding, and pushing against her belly, than to lie with a woman whose fetus lies facing her back, offering only its own back and its smooth rolling motions against her belly. There is an interaction between the fetus and these others, varying with how active a fetus is, how it lies, how it moves, and how the other responds, whether moving away repulsed, prodding back, crooning, laughing, leaning in for more, or pulling back from this intimacy.

A mother will use this movement of the fetus to create bonds between the fetus within and the rest of the family. How many women, how many times, have pulled the fathers' hands close to the belly and said, "Feel that." How many women have held children against them and said, "Feel the baby move." Also, some fathers reach out for this contact themselves. In his book, *The Nurturing Father*, Kyle D. Pruett describes his interview with Peter and Susan about their experience with pregnancy:

> ...he could not keep his hands off his wife's body, especially her swollen abdomen. "She was so beautiful and so huge." Once Susan felt the quickening of life within her, Peter would spend "what seemed like hours" with his hands on Susan's belly just "waiting to feel something happen." He became convinced that his unborn child was "tapping back messages in code"

when he tapped rhythmically on his wife's abdomen; "I could make him kick back, I really could." ...He even began singing to the fetus in the last trimester because Susan thought his voice "calmed the baby down. Probably me, too," she added later.

And then, holding his newborn son in his arms minutes after his birth:

> I remember when he kicked his brand new little feet against me as I held him in my arms that it felt almost familiar, like when we were tapping out code back and forth while he was still inside Susan.[7]

These relationships with babies begin to form before birth. Certainly they are relatively one-sided: the tapping had meaning for the father it could not have had for the baby. All relationships with babies, for quite some weeks or months after birth, are relatively one-sided. However, they are relationships, and they are beginnings of social interaction. Is a baby's "answering" kick *in utero* a "social" interaction on the baby's part? No more so than when, after birth, a baby will grasp tightly to an offered finger, but these reactions, social or not, call forth a wealth of social meanings. They are responded to socially—and that in turn calls forth more response. It takes years of this "calling forth," in Caroline Whitbeck's beautiful phrase[8], before a baby becomes a social creature. It goes on for years—for a lifetime—but it begins before birth.

Human beings are not chromosomes grown up—eggs and sperms writ large. We are essentially—in our very essence—social beings. We need to develop a technology for aiding procreation that reflects, and not destroys, our social connections.

References

[1] Oakley, A. (1984) *The Captured Womb: A History of the Medical Care of Pregnant Women* (Basil Blackwell Publisher Ltd., Oxford).

[2] Pritchard, J. A. and Macdonald, P. C. (1980) *William's Obstetrics,* 16th Edition (Appleton Century Crofts, New York).

[3] Brewer, G. S. (1977) *What Every Pregnant Woman Should Know: The Truth about Diets and Drugs in Pregnancy* (Random House, New York).

[4] Whitbeck, C. (1973) Theories of sex differences, in *The Philosophical Forum*, **VI**, 1–2.

[5] O'Brien, M. (1981) *The Politics of Reproduction* (Routledge and Kegan Paul, Ltd., London and Boston).

[6] Kitzinger, S. (1987) Getting in touch with your baby, in *Your Baby, Your Way: Making Pregnancy Decisions and Birth Plans* (Pantheon Books, New York).

[7] Pruett, K. D. (1987) *The Nurturing Father: Journey Toward the Complete Man* (Warner Books, New York).

[8] Whitbeck, C. (1981) Introductory remarks, in *The Custom Made Child?* (Holmes, H. B., Hoskins, B. B., and Gross, M., eds.), Humana Press, Clifton, N J.

Science, Conscience, and Public Policy

Historical Reflections on Controversial Reproductive Issues

John M. Eyler

A squat grey building of only thirty-four stories. Over the main entrance the words Central London Hatchery and Conditioning Centre, and, in a shield, the World State's motto, Community, Identity, Stability...

"And this," said the Director opening the door, "Is the Fertilizing Room."

Bent over their instruments, three hundred Fertilizers were plunged, as the Director of Hatcheries and Conditioning entered the room, in the scarcely breathing silence, the absent-minded, soliloquizing hum or whistle, of absorbed concentration. A troop of newly arrived students, very young, pink and callow, followed nervously, rather abjectly, at the Director's heels. Each of them carried a notebook, in which, whenever the great man spoke, he desperately scribbled. Straight from the horse's mouth. It was a rare privilege.[1]

—***Aldous Huxley**, Brave New World*

We follow these students and learn about the activities in this strange institution in the year After Ford 632. Future citizens, we discover, are conceived, gestated, and conditioned in vitro. Sperm and ova are carefully selected, using criteria determined by the Social Predestination Department. The upper castes, Alphas and Betas, are produced individually. Lower castes, Gammas, Deltas, and Epsilons, are mass-produced. To produce the nearly anonymous drones of society, each fertilized ovum of lower-class stock is subjected to the "Bokanovsky Process," causing it to bud. The result is as many as 96 identical embryos. Environmental and chemical manipulation during gestation, and behavioral modification and extensive use of hypnopedia (or sleep-teaching), after birth, produce citizens perfectly suited, both physically and psychologically, to their preordained places. Resentment, dissent, and fundamental questioning have become impossible. Even prejudices are state-selected and implanted. Community, Identity, and Stability have indeed been achieved.

Our group of students is reminded of some of the teachings of the state's founder, who seems to be an amalgam of Sigmund Freud and Henry Ford. Family life, when it existed, was a great source of conflict and unhappiness. History is bunk. The students make the sign of the "T" on their chests when certain of these truths are repeated. In this state, the family has been dissolved, and the past has been destroyed. The lower castes are even conditioned to have a fear and a revulsion of books. Freedom, autonomy, and civil rights have become irrelevant.

This is, of course, the opening of *Brave New World*, which was first published in 1932. In such purposeful intervention in human reproduction, Aldous Huxley gives us a terrifying picture of future technocratic, totalitarian states. The popularity of this novel suggests that many in his generation shared this anxiety. The 1930s were a period when attitudes and policies toward human reproduction were changing in Western nations, but it was certainly not the first time when reproductive policy was controversial. That decade, in fact, stands at the midpoint in lengthy discussions on whether and how technical and scientific intervention in human reproduction might be used. My purpose in this short essay is to review some of the better-known episodes in the history of these policy debates, to see what broad generalizations might be made about how our culture has dealt with these questions. We will consider in turn: abortion in the nineteenth century, contraception in the early twentieth century, and eugenics in the first four decades of the twentieth century.

Abortion in the Nineteenth Century

Let us begin by considering the abortion issue in America during the nineteenth century.[2,3] American law, like British common law, made taking the life of an unborn child a crime. We know that, despite this prohibition, abortions took place with some regularity in the eighteenth and early nineteenth centuries. Advice on how the mother herself might try to induce abortion was available in domestic manuals or from patent-medicine vendors. The reader might be warned that the medicine being discussed should not be taken during pregnancy, since it was sure to cause a miscarriage. The meaning of such warnings could not have been missed by many anxious readers. Abortion services were also provided by physicians and others, frequently under the guise of treating interrupted menstruation or other feminine disease. Such abortions were performed discreetly, but their existence could not have gone unnoticed by community leaders. Until roughly mid-nineteenth century, however, abortion did not cause much concern and was certainly not a public issue.

Indictments on abortion charges were uncommon and convictions exceedingly rare. Americans in the early nineteenth century assumed that the woman seeking an abortion was young, unmarried, and that she had been seduced. In other words, she was assumed to be a victim herself. Under these circumstances, legal remedies were seldom sought, and when they were, judges and juries were likely to show leniency. There were legal reasons as well for the rarity of convictions. The law recognized "quickening," the first sign of fetal movement, as the beginning of life. In practice, that standard meant that abortions taking place before quickening were not criminal. Since in the early stages of pregnancy the mother was the best judge of whether quickening had taken place, there was much opportunity for abortions to be sought and offered in the name of therapy.

One of the most surprising aspects of this episode is that this state of affairs continued long after the biological plausibility of the doctrine of quickening was exhausted. Only around 1860 did abortion become a political issue, and its politicizing resulted in a rush of new legislation. Thirty separate state abortion laws were passed between 1866 and 1877. By 1900, abortion statutes were in effect in every state except Kentucky. The drive for abortion legislation came initially from state and county medical societies. It was they who first forced the issue into the public arena. Scientific knowledge of reproduction, as imperfect as it was at mid-century, allowed physicians to argue with authority that the quickening doctrine was indefensible, and that abortions taking place before fetal movement was manifest were really no different from those taking place thereafter. They also emphasized the dangers that abortions posed to the mother's health and life. A curious feature of this early phase of the abortion debates was the role of two groups currently in the eye of the storm. First, American churches, much to the disgust of medical opponents of abortion, were slow to take a position. Second, nineteenth-century feminists, unlike their successors in the late twentieth century, usually supported antiabortion legislation.

The abortion policy in effect in the first half of the twentieth century was thus of fairly recent origin. Americans reformulated their attitudes and their laws toward abortion in the last third of the nineteenth century. Biological knowledge about the nature of gestation and fetal development were not enough in themselves to force such a reformulation. Social and cultural changes seem to have played a decisive role. Among these were changes in abortion practices themselves. Abortion was becoming commercialized. By the 1840s, New York City newspapers were carrying open advertisements from abortionists. A flourishing urban trade developed that attracted women from outlying areas. One abortionist, who assumed the name Madame Restell, became an infamous public figure in the middle decades of the century. She was a successful entrepreneur, with branch offices in Boston and Philadelphia and a staff of traveling representatives. Such a publicly conducted commercial enterprise was harder to ignore than the abortion practices of earlier decades. Furthermore, the stereotype of women who sought abortions also seems to have changed. They were now frequently portrayed as older, often married, and frequently middle class. That revelation was very significant to an age and a social class that were beginning to worry about the declining birth rates of white, native-born families. Abortion, in other words, began to seem both more common and more threatening than it had previously been. Under these circumstances, the biological arguments against the old social accommodation to abortion were more compelling, and medical opponents of abortion easily gained a sympathetic hearing.

Once abortion had been politicized, America moved quite suddenly from very lax to very strict standards for its tolerance. Consider the New York State statute of 1869, which served as a model for many other state laws. The act made abortion a felony at any period of gestation. Its terms applied to both the abortionist and the woman seeking an abortion. It applied to attempted abortion even when the woman was not pregnant. It also made it a crime to advertise or to offer abortion services. Abortion was legal only to save the mother's life. Not only were the new statutes strict, but they were enforced. Openly advertised abortion services were soon suppressed. Madame Restell herself was arrested in 1878. As if to symbolize the death of the old standards, she committed suicide the day before her trial was to begin.

Contraceptives in the Early Twentieth Century

This swift change in the law and the vigorous enforcement of the new statutes were results of the fact that the abortion issue was swept up in the Purity Crusade of the late nineteenth century,[4,5] a campaign against commercialized vice: pornography, prostitution, and abortion in particular. The activists were not just prudes and prigs. Among them were earnest men and women, deeply troubled by changes in American society and culture that accompanied urbanization: the loosening of family ties, the anonymity of city life, and the weakening of the moral code that had prevailed in smaller, more close-knit communities. The Purity Crusade inherited much of the moral zeal and many of the former leaders of the abolitionist campaign. It had the support of prominent philanthropists and social uplifters who saw in the moral transformation of the individual the means for reforming society. Although they emphasized moral persuasion, the purity crusaders were not above using state power to enforce their will.

Perhaps the best-remembered purity crusader is Anthony Comstock. With the help of other crusaders, Comstock lobbied an amendment of the postal code through Congress in 1873. The new provision tightened the prohibition on sending obscene material through the US mails, and it established a very broad definition of obscenity. The new standards of obscenity included any information or device for the control of conception. Thus, although the major nonsurgical techniques of contraception other than the birth control pill were available in America by 1870,[6] their use was actively discouraged. Comstock saw to it that the new code was vigorously enforced. He obtained an appointment as a special postal inspector, and in the first year traveled over 23,000 miles on his inspector's pass, made 55 arrests, and seized some 60,000 obscene rubber articles. For Comstock it was a holy war—a war whose ends justified the use of decoy letters, entrapment, and other suspect police measures.[7] The federal postal statute was soon supplemented by state laws making the offering of birth control information or devices illegal.

Comstock, of course, could not have acted with this authority and efficiency if he had not had important support. Many Americans in the last decades of the nineteenth century considered frank discussion of any sexual or reproductive matter socially taboo. In respectable middle-class circles, Comstock's goals, if not always his methods, may well have had wide support.[8,*] However, it is certain that the effects of this tough antiobscenity policy were felt far beyond the city streets. There is evidence that such policy inhibited medical teaching and discussion of sexuality in the late nineteenth and early twentieth centuries.[9] Certain sections of medical texts were removed to avoid offending the eyes of these guardians of public morality. Although some physicians had written in support of contraception and had recommended specific contraceptive techniques, most members of the medical profession opposed abortion and contraception, and shared prevailing standards of propriety. However, from the late 1890s, one can find doctors complaining about the poor training in human sexuality available to medical students.

Whereas it is plausible that purity crusaders spoke for majority public opinion in the 1880s, the laws and their strict application remained in effect long after that consensus had begun to change. Americans sought contraceptive information long before it was legal for them to obtain it. One might, in fact, interpret the first quarter century of Margaret Sanger's career as a contraceptive activist as a campaign to meet that demand by forcing the public, the courts, and the medical profession to see in Comstockery both a threat to civil liberties and an affront to the confidentiality of the doctor–patient relationship. It took two decades of civil disobedience and propaganda to open the mails to contraceptive information and devices intended for physicians' use only (*US v. One Package*,

---

*Major publishing houses adhered to a gentleman's agreement on good taste that made Comstock's intervention in creative literature unnecessary in the nineteenth century.

1937), to win the right to operate contraceptive clinics, and to gain professional approval for contraception as a legitimate medical service (AMA resolution of 1937).[10,11]

The Eugenics Movement

In the eugenics movement of the early twentieth century, we can see even more graphically how reproductive policies are the product of both scientific knowledge and technical capacity on the one hand, and social and cultural values on the other.[†] Eugenics, the program of improving human nature by controlling human reproduction, had scientific plausibility at the turn of the twentieth century. Humans, it was argued, should be subject to the laws of organic evolution that governed lower animals. There was evidence that certain human traits were inherited and that acquired characteristics were not. Finally, the rediscovery of Mendel's laws in 1900 seemed to promise understanding of how such inheritance took place and to make possible human intervention to achieve desired results. The prospect of using biological knowledge to improve the human condition was attractive to scientists, physicians, and legislators. The problem was that the applications proposed were grossly premature and the motives for intervention often suspect. Consider a defense of eugenics offered by Charles B. Davenport, a prominent biologist and one of the leaders of eugenics in America, in 1910:

> Society must protect itself...as it claims the right to deprive the murderer of his life so also it may annihi-

---

[†]For histories of the eugenics movement in America *see* Kenneth M. Ludmerer (1972) *Genetics and American Society: A Historical Appraisal*, Baltimore: The Johns Hopkins University Press; Mark M. Haller (1963) *Eugenics: Hereditarian Attitudes in American Thought*, New Brunswick, NJ: Rutgers University Press; and Garland E. Allen (1983) The misuse of biological hierarchies: The American eugenics movement, 1900–1940, *History & Philosophy of the Life Sciences* 5, 105–128.

late the hideous serpent of hopelessly vicious protoplasm.[12]

Eugenicists believed that such "hopelessly vicious protoplasm" exhibited itself in a host of human conditions: in mental retardation and feeblemindedness, in insanity, in physical handicaps and chronic diseases, in crime, in alcoholism and drug addiction, and, of course, in the nature of inferior races. They argued not only that such conditions were hereditary, but also that undesirable traits obeyed simple Mendelian laws of inheritance. They also argued that the degenerate human types who exhibited such undesirable traits were reproducing faster than better human stock and that, in the words of Theodore Roosevelt, America was committing "race suicide."

Eugenicists promised that, by applying the principles of inheritance, it would be possible to eliminate many of the most undesirable human traits in just a few generations. Also, they predicted enormous benefits to society in lower crime rates, in decreased institutional costs, and in improved racial vigor. As Albert E. Wiggam demonstrates, they proposed simple biological solutions for complex social problems.

> It [heredity and blood] has cost America a large share of its labor troubles, its political chaos, many of its frightful riots and bombings—the doings and undoings of its undesirable citizens. Investigation proves that an enormous proportion of its undesirable citizens are descended from undesirable blood overseas.[13]

Eugenic policy in the early twentieth century found two foci in America: state compulsory sterilization laws and immigration restriction. The first compulsory sterilization law was passed in Indiana in 1907.[14] By 1931, such laws had been passed in 30 states and were still in force in

27 of these. However, there is evidence that sterilization of institution inmates had begun before such legal authorization was available. For example, one of the chief proponents of the first legislation in Indiana, Dr. Henry C. Sharp, Surgeon of the Indiana Reformatory, had already begun sterilizing prisoners in his institution before the Indiana law was passed. He seems to have escaped criticism, perhaps because the vasectomies he performed stood in contrast to even harsher measures. In 1893, Dr. F. Hoyt Pilcher, Medical Director of the Kansas Asylum for Idiotic and Imbecile Youth, castrated 57 inmates. In the controversy that followed, he lost his job.[15] In most states, compulsory sterilization was not energetically pursued. In California, however, the superintendent of state hospitals, Dr. F. W. Hatch, was a sterilization activist. By 1935, California had sterilized some 12,000 inmates. That number represents about half of the total number then compulsorily sterilized in the entire nation.

The eugenicists did not originate the demand for limitation of foreign immigration, but their hereditarian arguments added weight to the economic and cultural justifications for such limitation in the aftermath of the First World War. The pressure of these combined forces proved effective. In 1924, Congress passed the Immigration Limitation Act (Johnson Act). This act set select immigration quotas by nation. Significantly, the 1890 census was used as the basis for computing the annual quota for each nation. This choice of reference line guaranteed that heavy immigration would cease from southern and eastern Europe, an area that the eugenicists claimed contributed many undesirable types.

If the program of American eugenicists had scientific plausibility in 1900, that plausibility soon evaporated in the light of new biological and anthropological studies.[16]

By the outbreak of the First World War, it was becoming clear that feeblemindedness, that convenient organic label eugenicists had used to explain so many social and psychological problems, was not a simple entity, but a complex of many disorders having a variety of causes. Scientists were also beginning to appreciate that the behavioral and personality phenomena that preoccupied the eugenicists were probably not determined by a single gene, and that their inheritance did not depend on simple Mendelian laws. The more complex model of human inheritance that was being constructed made it much more unlikely that any particular trait could be eliminated from a human population in the short periods of time the eugenicists promised.

As the scientific understanding of heredity began to change, scientists and physicians modified their attitudes toward eugenics. The British scientist J. B. S. Haldane used examples from America to warn his compatriots of the potential for abuse of eugenics. In his book *Heredity and Politics,* he describes the cases of two thieves who had been sterilized by court order in Yakima County, Washington. One of them, a poor white by the name of Hill, had stolen some hams to feed a hungry family. The other, Chris McCauley, was described by the judge in these terms:

> This man, about thirty-five years of age, is subnormal mentally and has every appearance and indication of immorality. He has a strain of Negro blood in his veins, and has a disgusting and lustful appearance.[17]

Haldane added:

> It is, I think, clear that Hill would not have been sterilized had he possessed an independent income. And

it is unlikely that McCauley would have been, had his complexion been lighter and his appearance more in conformity with Judge Holden's aesthetic standards. In my own judgement [sic] at least one well-known cinema "star" has a disgusting and lustful appearance, but I claim no scientific basis for this opinion, nor do I suggest that it be used as a basis for eugenical sterilization. In view of the judgements [sic] quoted it is of interest that British eugenicists often state that in America sterilization is carried out with a complete disregard for class or race distinctions.[18]

However, such direct, public repudiation of eugenics was rare before the Second World War. More typical responses were the withdrawal of support and the canceling of memberships and subscriptions. Although biologists were growing privately cynical and suspicious of eugenics in the interwar years, they did not take a public stand in opposition, and they left the field to the most dogmatic eugenic activitists. Even left-wing biologists, such as Haldane, who sometimes openly criticized eugenic programs, remained sympathetic to the claims of eugenics.[19] What Haldane opposed was what he considered the abuse of eugenics by capitalist nations to support racism and class divisions. To the end of his life, he remained a supporter of eugenic goals. It took the unfolding of Nazi population policies after 1933 to discredit the social analysis of eugenics and its political agenda. Recent historians have pointed out that the belief in biological determinism of complex intellectual and moral characteristics that buttressed eugenics has never been repudiated by biologists.[20] Rather, there has been a passing of scientific generations, and the older generation of eugenicists found it harder to recruit supporters in the changed social and political climate of the postwar era.

Perhaps we can understand why state control of human reproduction occupied so prominent a place in *Brave New World*. Huxley's generation was learning that technology could or soon might permit extensive control of reproduction, and that generation was beginning to contemplate the potential for abuse that lay in that direction. If America permitted compulsory sterilization, what might the new totalitarian regimes prove capable of doing?

Our brief survey of the histories of several controversial reproductive issues suggests that our culture has dealt with such issues in a desultory manner. Our positions have been determined by a complex mixture of knowledge, values, and prejudice. On one hand, we have been slow to recognize some issues, such as the use of abortion as a method of birth control in the nineteenth century, and slow to modify our established ways of reconciling ideals with reality, such as the legal use of the doctrine of quickening, in the light of scientific knowledge. On the other hand, when we have acted to formulate new policies, such as the criminalization of abortion and contraception, we have often done so categorically, often punitively, producing rigid standards that were difficult to modify as our values changed. Finally, we have sometimes applied scientific knowledge prematurely and have let the authority of science mask other social and political agenda. The eugenics movement of the twentieth century stands as the most grim warning of the dangers that lie in the misuse of the power to control human reproduction.

References

[1]Huxley, A. (1970) *Brave New World* (Chatto and Windus, London).

[2]Mohr, J. C. (1978) *Abortion in America: The Origins and Evolution of National Policy, 1800–1900* (Oxford University Press, New York and Oxford).

[3]Smith-Rosenburg, C. (1985) The abortion movement and the AMA, 1850–1880, in *Disorderly Conduct: Visions of Gender in Victorian America* (Oxford University Press, New York and Oxford).

[4]Pivar, D. J. (1973) *Purity Crusade, Sexual Morality and Social Control, 1868–1890* (Greenwood Press, Westport, CT).

[5]Reed, J. (1984) *The Birth Control Movement and American Society: From A Private Vice to Public Virtue* (Princeton University Press, Princeton, NJ).

[6]Reed, J.

[7]Reed, J.

[8]Boyer, P. S. (1968) *Purity in Print: The Vice-Society Movement and Book Censorship in America* (Charles Scribner's Sons, New York).

[9]Reed, J.

[10]Reed, J.

[11]Kennedy, D. M. (1970) *Birth Control of America: The Career of Margaret Sanger* (Yale University Press, New Haven and London).

[12]Davenport, C. B. (1910) Report of Committee on Eugenics—*American Breeders' Magazine* 1, in Ludmerer 8.

[13]Wiggam, A. E. (1924) *The Fruit of the Family Tree* (Bobbs-Merrill Co., Indianapolis, IN).

[14]Reilly, P. (1983) The surgical solution: The writings of activist physicians in the early days of eugenical sterilization. *Perspectives in Biology & Medicine* 26, 648.

[15]Reilly, P.

[16]Ludmerer, K. M. and Allen, G. E. (Reference in Note 2)

[17]Judge Holden cited in Haldane, J. B. S. (1938) *Heredity and Politics* (W.W. Norton and Company, New York), pp. 104, 105.

[18]Haldane, J. B. S.

[19]Paul, D. (1984) Eugenics and the left. *Journal of the History of Ideas* 45, 567–590.

[20]Paul, D.

Current Religious Perspectives on the New Reproductive Techniques

Baruch Brody

The publication of the Office of Technology Assessment's major report entitled *Infertility: Medical and Social Choices*[1] constitutes a major contribution to our understanding of the many issues surrounding the treatment of infertility. Of special significance is its treatment of the ethical and religious considerations relevant to the new reproductive techniques. In this essay, I want to elaborate upon that treatment. My goals are to identify the major concerns about these techniques shared by many of America's religious communities and to discuss the implications of the existence of these concerns for the formulation of public policy in this area.

There are two preliminary points that need to be carefully noted. To begin with, it is very unhelpful to think of religious Americans as being divided into Catholics, Protestants, and Jews. After all, America contains other religious communities (e.g., Eastern Orthodox, Muslim, Mormons). Moreover, Protestants and Jews are nonhomogeneous communities when it comes to the formulation of moral attitudes. The contrasts between Episcopalians and Evangelicals and between Orthodox Jews and Reform

Jews need to be kept in mind. Hence, when I talk about America's religious communities, I am talking about a large number of different groups. Secondly, it is important to remember that there is often a difference between the official teachings of a particular religious community (in the case of those many communities that have official teachings) and the beliefs of individuals who belong to those communities. Individual members of a particular community may be unaware of the community's official teachings on a particular topic, or they may be aware of it and disagree with it. Think, for example, of the many Catholics who disagree with their church's teachings on artificial contraception. When I talk about the concerns of America's religious communities regarding the new reproductive techniques, I am talking about the concerns expressed in their official teachings.

This essay will examine the concerns expressed in the official teachings of a large number of America's religious communities about aspects of the new reproductive techniques. The first section will identify those concerns. The second section will document their presence in the various communities. The third section will discuss the policy implications of the existence of these concerns.

Concerns of Religious Communities

There are at least six major concerns that have been expressed by America's religious communities about the new reproductive techniques. They can be summarized as follows:

1. The new reproductive techniques disrupt the connection between unitive conjugal intimacy and procreative potential that is required by morality.

Religious Perspectives 47

2. The new reproductive techniques often introduce third parties into the process of reproduction, and this is morally illicit.

3. The new reproductive techniques often result in a morally illicit confusion of lineage, since children are unaware of their biological parents.

4. Some new reproductive techniques (IVF) often involve a failure to implant fertilized eggs. This is a form of early abortion and is therefore morally illicit.

5. The new reproductive techniques often involve a dehumanization of the reproductive process and are therefore morally illicit.

6. Some new reproductive techniques (especially surrogacy) involve commercialization and exploitation that make them illicit.

Each of these concerns needs to be explained and developed. Before doing so, however, there is an important point that needs to be made about all six of these concerns. None of them involves the claim that the rights of anyone are being violated by the use of the new reproductive techniques. None of them involves the claim that some individual is being harmed by the use of the new reproductive techniques. None of them involves the claim that the use of the new reproductive techniques constitutes an injustice to anyone. In short, none of them involves an appeal to the principles of rights, of consequences, and of justice that dominate secular discussions of bioethics. This might lead some readers to conclude that these are not really moral concerns, and that these are at most theological concerns. Such a conclusion would be a mistake; it would rest upon a much too narrow conception of morality. These religious concerns involve appeals to the portion of morality that involves deontological constraints. These concerns appeal to the claim that certain types of actions are intrinsically wrong, independent of any consequences they have, independent of any violations of rights they involve, and independent of any injustices they produce. It is an important feature of most religious moralities—but not just of religious moralities—that they involve these deontological constraints on morally licit actions. With this understanding in mind, we turn to an examination of each of these concerns.

The first of these concerns is based upon a special understanding of human sexuality and of human reproduction. It comes in two major versions. One is the claim that each licit human sexual act must involve two different aspects joined together, unitive conjugal intimacy and reproductive potential. Such a claim has profound implications, both for sexuality and for reproduction. It entails that particular sexual acts that lack reproductive potential (e.g., masturbation and coital sexual acts employing artificial contraceptive techniques) are morally illicit. It also entails that particular acts of reproduction that do not involve unitive conjugal intimacy (e.g., artificial insemination and reproductive acts not involving a married couple) are morally illicit. The other version claims that this link between unitive conjugal intimacy and reproductive potential must be found in every morally licit sexual relation, but it does not need to be found in each particular sexual act. Individual acts of conjugal sexuality employing artificial contraceptive techniques are morally licit according to this second version, but a lifetime of conjugal sexuality employing contraception—because the couple wishes to have no children—is morally illicit. Artificial insemination using the sperm of a husband to inseminate his wife is morally licit according to this second version (because the reproductive potential grows out of a relation involving unitive conjugal intimacy), but the artificial insemination of a single woman is morally illicit (because the reproductive potential is totally divorced from any unitive conjugal intimacy).

The second concern relates to those reproductive techniques (artificial insemination or in vitro fertilization using donor sperm) in which gametes from a party not married to the woman in question are introduced into the reproductive process. The claim of this concern is that it is this introduction that makes the reproductive process illicit. Again, this concern comes in two major versions. One claims that such reproductive acts are either acts of adultery or acts of fornication, and are as illicit as any other act of adultery or fornication. To be sure, these acts do not involve the component of sexual pleasure found in more traditional versions of these illicit sexual acts, and, if the spouse consents, they also do not involve any betrayal. According to this first version, however, these differences are not morally relevant. The second version of this concern disagrees. It does not consider these acts to be acts of adultery or acts of fornication; however, it still considers them to be illicit.

The third concern relates to those reproductive techniques (artificial insemination or in vitro fertilization using donor sperm) in which at least one of the parents who raise the child is not the biological parent of the child and in which the child does not know his or her biological parent. In these techniques, knowledge of the child's biological lineage is destroyed. It is just this aspect of these techniques that raises the third concern. Those voicing this third concern think that knowledge of biological lineage is morally significant, and that it is morally wrong to use techniques that destroy it. They may or may not want to argue that it is also psychologically damaging to the child; their moral claim is independent of that psychological claim. The moral claim is based upon a moral view of the importance of known lineage. Sometimes, it is claimed that knowledge of lineage is important to avoid unintended incestuous relations. On other occasions, it is claimed that knowledge of lineage is morally important even independent of that consideration. Those who raise this third concern must, it should be noted, also oppose adoption as it has been practiced in the West in the recent past. Although they can certainly approve of people raising orphans, they must disapprove of any attempt to hide the biological lineage of the orphan.

The fourth concern relates to a special problem faced by in vitro fertilization programs. For a variety of technical reasons, many eggs are retrieved in a given cycle. If too many fertilizations result, some of the fertilized eggs will not be transferred back to the woman. Until recently, the nontransferred fertilized eggs were destroyed. Now, attempts are being made to freeze them and to use them in future cycles. This fourth concern challenges the destruction of the nontransferred fertilized eggs on the grounds that this destruction constitutes an illicit early abortion. Naturally, those who raise this concern must view abortions as illicit from the moment of conception; those whose opposition to abortion is an opposition to abortions from implantation (or sometime later) will not be troubled by this concern. Moreover, this concern can be met by those programs that limit the number of eggs they attempt to fertilize or that transfer all fertilized eggs back to the woman. Still, given the practice of many in vitro programs, this is a real concern for those communities who oppose abortion from the moment of conception.

The fifth concern relates to the technological feature of the new reproductive techniques. Especially in cases of in vitro fertilization, but even in artificial insemination, the human act of reproduction is transformed into a technological achievement. Those who raise this fifth concern

Religious Perspectives 51

view this as a morally illicit dehumanization of the reproductive process. Some who raise this concern would object to this dehumanization only if the new techniques were used as a substitute for reproduction growing out of conjugal sexuality; they would accept the use of such techniques, however, as a measure employed by otherwise infertile married couples. Others who raise this concern claim, however, that the need to resist dehumanization is so great that such technologically based reproductions should be eschewed in all cases.

The last of these concerns is directed primarily against surrogacy. It is rooted in the fact that most surrogacy arrangements involve a payment to the surrogate mother over and above the reimbursement of any expenses she incurs. This commercialization of reproduction is viewed as objectionable for two reasons. To begin with, it is believed that the women in question are being exploited. It is important to keep in mind that exploitation is not the same thing as coercion, even if both of these difficult-to-define concepts are closely related. The women who serve as surrogate mothers often volunteer; it is difficult, therefore, to see them as being coerced into serving as surrogate mothers. One might nevertheless claim that their economic need leads them to volunteer, and that the infertile couples are exploiting them. This is the first reason offered to oppose commercial surrogacy. The second reason is that the commercialization is illicit even if it is not coercive or exploitative. Those who offer this objection might compare commercial surrogacy to the sale of body parts (e.g., second kidneys). Their point is that some things just should not be commercialized.

These, then, are the six major concerns raised by America's religious communities about the new reproductive techniques. It is clear that these concerns do not apply equally to all of the new techniques. For example, very few of them can be raised against artificial insemination using the husband's sperm. The only ones that are applicable are the first concern (and only the version that insists that each licit sexual *act* must combine reproductive potential and unitive conjugal intimacy) and the concern about dehumanization. In vitro fertilization using gametes from a married couple raises no additional concerns unless there is some wastage of fertilized eggs; if there is, then the concern about early abortions arises. Artificial insemination or in vitro fertilization using donor sperm is very different. All but the last concern challenge the moral licitness of those techniques. Finally, all six concerns arise in cases of commercial surrogacy, since they are also cases of artificial insemination using donor sperm.

Concerns of Specific Religious Communities

Having presented the set of concerns that have been raised about the new reproductive techniques by America's religious communities, I shall now document the presence in the various religious communities of these concerns. In order to do that, however, I must first say something about the nature of authority in the various religious communities. After all, different communities have very different theories about what constitutes the authoritative official teachings of their community. If our goal is to study the official teachings of America's religious communities so as to discover the extent to which they share these concerns about the new reproductive techniques, then we need to take care that the documents we examine actually reflect authoritative official teachings in light of that community's theory about what the sources of authority are within that community.

America's religious communities differ greatly on the question of authority. Their views range from the claim that there is no authority other than the individual's understanding of scripture (e.g., Baptists) to the claim that there is a single individual whose teachings are authoritative for the entire community (e.g., Roman Catholics). In between these two extremes are communities that vest authority in regular councils of their leadership (e.g., Episcopalians and Presbyterians), communities that vest authority for individual subcommunities in the subcommunity's consecrated clerical leader (e.g., Greek Orthodox), and communities that vest authority in the hands of recognized teachers (e.g., Orthodox Jews). Many other variations and combinations are possible. Moreover, communities—even the ones listed above—have unofficial sources of authoritative teachings as well. This is not the place, however, to explore all of these complexities. It is sufficient for our purpose to note these possibilities and to be sensitive to them as we gather our evidence from the literature of various religious communities.

Which communities share the first of our concerns, the concern that the new reproductive techniques disrupt the connection between unitive conjugal intimacy and procreative potential? That concern exists in two versions. The first insists that the connections must be present in each licit sexual act. That is certainly the official teaching of the Roman Catholic community. To quote a recent document issued by the Congregation for the Doctrine of the Faith and approved by Pope John Paul II:

> The Church's teaching on marriage and human procreation affirms the inseparable connection, willed by God and unable to be broken by man on his own initiative, between the two meanings of the conjugal act: the unitive meaning and the procreative mean-

> ing...Contraception deliberately deprives the conjugal act of its openness to procreation and in this way brings about a voluntary dissociation of the ends of marriage. Homologous artificial fertilization, in seeking a procreation which is not the fruit of *a specific act* of conjugal union, objectively effects an analogous separation between the goods and the meanings of marriage[2] [italics added].

I know of no other religious community that shares this view, but there are others that share this concern in its second version, according to which the connection must be present through the whole of a relation. For example, in approving of artificial insemination using the husband's but not a donor's sperm, the Commission on Theology and Church Relations of the Lutheran Church—Missouri Synod wrote:

> Here artificial insemination is offered as an aid to procreation within marriage. It is intended not to separate procreation from *the context of* the loving union of husband and wife. Instead, it is a way of bringing their love to the fruition towards which it is naturally ordered[3] [italics added].

Which communities share the second of our concerns—the concern that some of the techniques illicitly introduce third parties into the process of reproduction? Many do. Some of them, as noted above, treat the introduction of gametes from a third party (e.g., artificial insemination using donor sperm) as adultery or fornication. Thus, to quote an article from the Jehovah's Witnesses newspaper, *The Watchtower*:

> In this case the sperm and the egg come from husband and wife. This is noteworthy from the biblical standpoint. Why? Because of a law God gave to the ancient Israelites: "You must not give your emission as semen to the wife of your associate to become unclean

by it" (Lev. 18:20, 29)... From the Bible we must conclude that if conception is accomplished with the sperm and an egg not from husband and wife, it would amount to adultery or fornication (June 1, 1981, p. 31).

Others, who do not see it as adultery or fornication, nevertheless object to the use of gametes from someone other than the married couple. Thus, to quote from a split Anglican task force on this topic:

> We believe that A.I.D. given to a wife by a qualified physician, with the consent of her husband, should not be considered in either morals or law to be adultery...Some of us have serious personal reservations about A.I.D. based on our understanding of Christian teachings about marriage and about human nature...We believe that, given the psychosomatic unity of persons, the creation of new life should be by a husband and wife in the personal intimacy of sexual union.[4]

In between these two positions is the position of the Greek Orthodox church:

> This means that the egg must come from the wife's own ovaries, and that the sperm must be the husband's own; for a donor, whether male or female, would constitute intrusion of a third party into the marriage tantamount to adultery.[5]

Which communities are concerned about the destruction of known lineage in many forms of the new reproductive techniques? Most prominent are the Islamic community and portions of the Jewish Community. To quote from a well-known text from the Islamic world:

> Islam safeguards lineage by prohibiting Zina (adultery) and legal adoption, thus keeping the family line unambiguously defined without any foreign element

entering into it. It likewise prohibits what is known as artificial insemination if the donor of the semen is other than the husband.[6]

A very similar viewpoint is expressed in the following passage from a well-known guide to the rabbinic literature on bioethics:

> Concealment of paternal identity is unconscionable according to Jewish law. Halakhah requires a three-month waiting period following divorce or the death of a husband before remarriage is permitted...Rashi explains that the divine presence rests only upon those whose genealogy is clearly known. Moreover, the Talmud declares, there is concern lest the child marry a potential sibling...[7]

The fourth concern, about early abortions, is articulated most clearly in the Roman Catholic and Eastern Orthodox literature. Other communities holding similar views are the Jehovah's Witnesses and the Mormons. Since it is always a pleasure to show unity among communities that have a long history of strife, I illustrate the presence of this concern by quotations drawn from the Roman Catholic and the Eastern Orthodox community. The Catholic claim is the following:

> Human embryos obtained in vitro are human beings and subjects with rights; their dignity and right to life must be respected from the first moment of their existence. In the usual practice of in vitro fertilization, not all of the embryos are transferred to the woman's body; some are destroyed. Just as the church condemns induced abortion, so she also forbids acts against the life of these human beings.[8]

The Eastern Orthodox claim is this:

> Serious objection is raised here to the fact that many more eggs are fertilized than can be used; those not

Religious Perspectives

used are discarded. This is easily seen to be the killing of a potential life: abortion.[9]

The source of the fifth concern seems to be the writing of Leon Kass, a well-known bioethicist writing independently of any of the major religious traditions.[10] It has been particularly influential in the Eastern Orthodox tradition:

> Finally, objections must be raised in terms of the mentality created by such a practice. As a step which dehumanizes life and which separates so dramatically the personal relations of a married couple from child bearing it is very suspect.[11]

Since surrogacy has only recently emerged, it has not yet attracted the number of studies and policy statements that other reproductive techniques have attracted. It is nevertheless possible to see the sixth set of concerns emerging in many religious communities. A clear statement of the concern about commercialization is found in the following quotation from the British Council of Churches:

> Surrogate motherhood could conceivably be undertaken freely as a gift and service to a childless couple. In fact, however, it is much more likely to be a business transaction in which substantial sums of money change hands...Once this business approach has been introduced, what is to prevent its extension to the sale of babies, whatever their mode of production? Procreation is not a business. Babies are not to be treated as commodities.[12]

Similar statements have been put forward in the American context.

The concerns we have identified are thus widespread in America's religious communities. The documentation we have given here is only part of the evidence of that widespread concern. What, however, are the implications of these widespread concerns for social policy? That is the question we shall examine in the last part of this paper.

Policy Implications

In this final section of the paper, I turn from an analysis of a set of concerns and a documentation of their presence to an examination of the policy implications of these concerns. My goal is to raise questions rather than to attempt to provide answers, since the issues I want to raise are relatively new, and it would be premature to draw any conclusions until a great deal of analysis has been devoted to them.

There has been considerable discussion in the history of Western thought about what ought to be the relation between law and social policy on the one hand and ethics (secular and religious) on the other hand. This is, after all, the fundamental dispute between natural-law theorists and legal positivists. Connected with this great debate is the additional issue, raised so clearly by J. S. Mill in the nineteenth century in his classic *On Liberty* and discussed extensively since then,[13] of whether criminal law and other forms of social control should be used to control all behavior judged immoral, or whether they should be used to control just behavior that harms and/or infringes upon the rights of nonconsenting victims. This traditional debate has, of course, some relevance to our discussion. One could imagine members of those communities that judge forms of the new reproductive techniques to be immoral calling for a legal ban on the use of those techniques. In fact, the Catholic leadership has called for at least some of that in the following passage:

> The political authority is bound to guarantee to the institution of the family, upon which society is based, the judicial protection to which it has a right. From the very fact that it is at the service of people, the political authority must also be at the service of the family. Civil law (the law of the state) cannot grant approval to techniques of artificial procreation which, for the benefit of third parties (doctors, biologists, economic or governmental powers), take away what is a right inherent in the relation between spouses; and therefore civil law cannot legalize the donation of gametes between persons who are not legitimately united in marriage. Legislation must also prohibit, by virtue of the support which is due to the family, embryo banks, post mortem insemination, and "surrogate motherhood."[14]

However, it is not these issues that I wish to raise in this last section. I want to raise, instead, a different set of issues concerning the relation between law and public policy on the one hand and ethical concerns on the other hand.

I shall begin by listing a representative (but by no means complete) set of legal and/or policy issues that our society must face in relation to the new reproductive techniques:

1. Should the provision of these reproductive techniques to those who cannot afford to pay for them from their own funds be covered by public health insurance programs, such as Medicaid, the VA and the CHAMPUS programs, and municipal and county public hospitals?

2. Should we encourage the provision of these services by mandating that private health insurance polices that cover pregnancy-related services should also cover the provision of these new reproductive techniques, and by giving research to improve their effectiveness and lower

their cost a high priority in the public funding of biomedical research?

3. Should we continue to provide or commence to provide legal frameworks to deal with the many specific problems (e.g., legitimacy and parenthood, or the enforcement of agreements) raised by these techniques?

I shall now review each of these issues separately, in order to raise my questions about how these policy issues should be affected by the presence of these religiously rooted moral concerns.

One of the main problems of contemporary health policy is how to ensure access to health care for those who cannot afford to pay on their own (or with the help of private insurance) for the health care from which they can benefit. In a world of limited resources, we may be required as a matter of policy to limit the care provided to such people through public programs by prioritizing forms of health care and by providing only those forms that have high priority. One radical solution, which I have advocated in several places,[15,16] is to allow the recipients to do the prioritizing themselves. Most social programs are not structured this way; the providers do the prioritizing. It is with this background in mind that I raise the first of my questions about the relation between social policy and religiously rooted moral concern: Is the fact that some reproductive techniques are viewed by many citizens as morally illicit a good reason for giving the provision of those techniques a low priority for funding in a system of health care that is funded by taxes raised from all citizens, including the opponents? On the one hand, one might argue that doing so would be letting the opponents indirectly impose their values on the recipients, since the recipients have no other way of getting the care. On the other hand, one might argue that doing so is a way of respecting the

Religious Perspectives

values of the offended taxpayers, by making the fact that they find these forms of care morally offensive a reason for giving them a low priority.

One of the main obstacles to the spread of the new reproductive techniques—especially in vitro fertilization—is the very high cost of their use combined with the related fact that they are not covered by many private health insurance schemes. It has been proposed that insurance schemes covering pregnancy-related expenses should be mandated to cover infertility-related expenses.[17] It has also been suggested that priority should be given to funding research to improve the effectiveness and lower the cost of these techniques.[18] The idea behind these proposals is twofold. To begin with, they will promote the availability of those techniques. Secondly, they treat the fertile and the infertile equally, providing needed services and research for both. The second of my questions about the relation between social policy and religiously rooted moral concerns is this: In setting social policies to encourage the spread of various techniques, is the fact that the technique in question is viewed by many citizens as morally illicit a good reason for not making its spread a social goal? On the one hand, one might argue that we cannot let the religiously based moral beliefs of some citizens guide social policies. On the other hand, one might argue that so structuring social policy is a proper way of differentiating what ought to be a social goal from what ought only to be the goal of subgroups in society, viz., those that strongly approve of the technique in question.

One of the most controversial set of issues surrounding the new reproductive techniques is the need for legal decisions about their implications. From the time when donor gametes were first used in artificial insemination programs, societies have had to grapple with the question of the paternity of the children produced. With the development of the in vitro programs (including frozen embryos) and surrogacy programs, new legal questions have arisen: Who owns the not yet used fertilized embryos and can decide when and how they should be used? Can someone back out of a surrogacy contract and keep the child they have borne? All of this leads to my third set of questions about the relation between legal and social policy on the one hand and religiously rooted moral concerns on the other hand: Is the fact that some of these techniques are viewed by many citizens as morally illicit a reason for not passing legislation to deal with their implications? On the one hand, one might argue that, since these techniques are being employed, it is foolish not to settle in some systematic fashion the many questions raised by their use. On the other hand, one might argue that legislation about implications is—or is perceived as—an implicit social acceptance and social endorsement of the techniques in question, and it is wrong for society to give these techniques such an implicit endorsement.

What is emerging is relatively straightforward. All of our questions are really questions concerning how a pluralistic society should deal with the moral disagreements among its citizens when setting social policies. It is easy enough to say that it should try to adopt a neutral stance, but our questions show that neutrality is not so easy to define. Is funding the services for people who want them the neutral approach, or is treating such services as private matters to be funded privately the more neutral approach? Is encouraging their availability in an affordable fashion for those who want them the neutral approach, or is neither encouraging nor discouraging their availability the more neutral approach? Is legislating solutions to the many questions raised by their use the neutral approach, or is adopting a "hands-off" approach the more neutral approach? I don't know the answer to these questions. I don't think that we have, as yet, a good theory to answer these questions. They, to my mind, are the really important questions raised by the current religious perspectives on the new reproductive techniques.

References

[1] US Congress, Office of Technology Assessment (May 1988) *Infertility: Medical and Social Choices* (Government Printing Office, Washington, DC).

[2] Congregation for the Doctrine of the Faith (1987) *Instruction on Respect for Human Life* (Congregation, Vatican City).

[3] Commission on Theology and Church Relations (1981) *Human Sexuality: A Theological Perspective* (Lutheran ChurchMissouri Synod, St. Louis).

[4] Creighton, P. (ed.) (1977) *Artificial Insemination by Donor* (Anglican Book Center, Toronto).

[5] Department of Church and Society (n.d.) *Statements on Moral and Social Concerns* (Greek Orthodox Archdiocese, New York).

[6] Al-Qaradawi, Y. (n.d.) *The Lawful and the Prohibited in Islam* (American Trust Publications, Indianapolis).

[7] Bleich, D. (1981) *Judaism and Healing* (Ktav, New York).

[8] Congregation for the Doctrine of the Faith.

[9] Department of Church and Society.

[10] Kass, L. (1985) *Toward a More Natural Science* (Free Press, New York).

[11] Department of Church and Society.

[12] Free Church Federal Council and British Council of Churches (1982) *Choices in Childlessness* (Free Church Federal Council, London).

[13] Feinberg, J. (1984–6) *The Moral Limits of the Criminal Law*, Vols. 1–3 (Oxford University Press, New York).

[14] Congregation for the Doctrine of the Faith.

[15] Brody, B. (1987) Justice and Competitive Markets. *Journal of Medicine and Philosophy* 12, 37–50.

[16] Brody, B. (1981) Health Care for the Haves and Have-nots, in *Justice and Health Care* (Shelp, E., ed.), Reidel Publishing Company, Dordrecht, Holland.

[17] Office of Technology Assessment.

[18] Office of Technology Assessment.

Essential Ethical Considerations for Public Policy on Assisted Reproduction

Carol Tauer

The title of this chapter, "Essential Ethical Considerations" rather than "*The* Essential Ethical Considerations," indicates that there is no single, monolithic set of ethical considerations. Different religious groups have their own sets of ethical criteria, as do various cultural groupings. Even individuals differ as to the ethical concerns they view as important or even relevant. Every ethical issue can be examined from a variety of perspectives, each with its own set of ethical considerations.

In a systematic analysis of an ethical issue, a clear identification of perspective eliminates much ambiguity and confusion. One may begin with the question, "From whose perspective is this problem viewed?" or more concretely, "Who is the decision maker in this situation?" In many cases, the identification of the perspective or the decision maker is a matter of choice, even of arbitrary choice. However, unless a choice is made and some distinct perspective is taken, the analysis will be confused and confusing.

For reproductive techniques, the perspective chosen could be that of a reproductive biologist, a clinical physician, an infertile couple, a potential surrogate mother, a potential gamete donor, or one of the levels or branches of government. (Although an embryo, fetus, infant, or child may be the most critically affected in these decisions, none is a decision maker.) Moreover, the perspective taken can be that of a particular individual or of society as a whole.

An individual making an ethical decision is usually confronted by a different set of ethical considerations from those that society must weigh in looking at the same issue. For example, there may be nothing wrong with an individual's gambling for entertainment, but from society's perspective, there may be good ethical reasons for outlawing at least some forms of gambling. Another example, suggested by St. Thomas Aquinas: Although prostitution may be an immoral practice, there may also be good reasons for society's tolerating the existence of this practice.

This paper will focus on society as the decision maker and discuss the ethical considerations relevant to the development of public policy. A long tradition in political philosophy suggests that the following criteria should be considered in the establishment of public policy on moral issues. The criteria themselves can be regarded as ethical requirements for public policy decision making[1]:

1. The issue in question must impinge on the good of society, not concern merely the regulation of individual moral behavior.

2. Before adoption of a policy on the issue, there should be adequate opportunity for public debate.

3. Public debate should indicate a broad societal consensus on the proposed policy. (An issue where fundamental human rights are at stake would be an exception to this criterion. Persons do not gain and lose human rights on the basis of societal consensus, and a majority cannot ethically deprive minority group members of these rights.)

4. The proposed policy should be consistent with existing laws and regulations, or indicate that these should be changed.

5. After adoption, the policy should be promulgated in a way that can be understood by the public.

6. The policy should be enforceable and be enforced.

Fundamental Ethical Principles

Two government commissions, the National Commission for the Protection of Human Subjects[2] and the President's Commission for the Study of Ethical Problems in Medicine and Biomedical and Behavioral Research,[3] identified three basic ethical principles to guide the formation of public policy: respect for persons, well-being, and equity. The principle of respect for persons requires that we allow competent persons to act on their autonomous choices and that we protect incompetent persons where necessary. It also requires that we not use people as means to our ends, but respect them as ends in themselves. The principle of well-being requires that we promote the welfare of others and protect people from risks of harm. The principle of equity or justice requires that we treat persons fairly, avoiding discrimination as well as an inequitable distribution of social benefits and burdens.

Both commissions consciously rejected an overtly utilitarian stance whereby individuals might be utilized for the benefit of society. Rather, the common good is sought through promoting the good of individuals, and through insisting on equity or a fair distribution of societal resources.

These three basic principles have served us well for a number of years. As principles, however, they are broad, general, and abstract. Application to practical, concrete situations requires interpretation. In the area of assisted reproduction, the application of basic principles is not always immediately evident, and reasonable people may differ as to their interpretation. For example, the analytical approach taken by Michael Bayles[4] contrasts with the feminist approach of Christine Overall[5] and the authors represented in Michelle Stanworth's collection.[6] Differences at the international level are shown in the various committee statements juxtaposed by LeRoy Walters,[7] and in individual documents like that of the Council for Science and Society.[8] Even within a single religious body, such as the Roman Catholic Church, divergent interpretations are expressed.[9-11]

In my discussion, I will suggest possible interpretations of the three basic principles when applied to the new reproductive techniques. In suggesting these interpretations, I will highlight the significant ethical questions that must be resolved in relation to public policy.

Application of Principles to Assisted Reproduction

Reproductive techniques fall into two broad categories. The first consists of techniques performed in a scientific laboratory. In vitro fertilization is the laboratory procedure that is used most frequently and that raises the most ethical questions. The second consists of reproductive techniques in which the involvement of a third party,

External Fertilization

Process and Problems

In vitro fertilization—the creation of human life outside a woman's body—is sometimes called external fertilization. This language suggests the unusual situation to which it gives rise: the presence of early human embryos in the laboratory enables scientists to perform research and manipulation not possible otherwise.[12]

Although US clinics performing in vitro fertilization are focused almost entirely on providing babies to infertile couples, there is also a potential for new types of biological research. If, after fertilizing multiple oocytes, a laboratory confronts what are sometimes called "spare" embryos, what is to be done with them? Should they be frozen? Must freezing be followed by future transfer? Can they be used for research? On the other hand, perhaps all embryos should be returned to the egg donor, giving them a chance at life, but putting the woman at risk of a multiple pregnancy. What if some fertilized eggs do not look viable, and hence, would be poor candidates for transfer to the woman? Could these be used for research, rather than simply wasted? Then, if research involving early embryos is found to be acceptable, what about creating human embryos purely for research purposes?

According to many scientists, embryo research is an invaluable resource in investigations of genetic diseases,

---

*The two categories of assisted reproduction may, of course, overlap or be combined in practice. For example, in vitro fertilization can be combined with a surrogate arrangement through implantation of the cleaving embryo in a woman other than the oocyte donor. The separation of techniques into two categories is a conceptual division for purposes of ethical analysis.

the differentiation process, reasons for miscarriage, and even some forms of cancer. Although much basic research can be performed using animals and animal embryos, some aspects of genetic and reproductive biology require human material for study. Human in vitro fertilization has had little systematic investigation since its first successes.[13] In many clinics, the success rate for embryo transfers resulting in live births is extremely low; studies utilizing laboratory-fertilized embryos could provide the key to more successful clinical results.

Public Policy

The presence of early human embryos in the laboratory offers an opportunity for unique scientific advances, yet challenges personal, religious, and societal norms on the treatment of potential human life. These challenges suggest that our body politic ought to set standards for this research, as it does for other research involving human "subjects."

Federal regulations on fetal experimentation, implemented in 1975, require that federally funded research involving in vitro fertilization be reviewed at the national level by an Ethics Advisory Board (EAB) (45 CFR 46.204).[14] This board was appointed in 1977 by the Department of Health, Education, and Welfare (HEW), and was charged with examining the ethical acceptability of in vitro fertilization and related research. In particular, it was to give an ethical assessment of a research proposal from Pierre Soupart, already approved by NIH on its scientific merits, to study whether the in vitro process was safe for the embryos involved.

In its report, the EAB carefully applied ethical principles, set broad ethical criteria for in vitro fertilization

Ethical Considerations

practice and research,[†] and affirmed the ethical acceptability of the Soupart proposal.[15] In response, HEW tabled the Board's report, took no action on it, and essentially closed public policy debate on this matter. Soupart never received funding, nor has any human in vitro research been considered for federal funding since then. In addition, the EAB was eliminated when its charter and funding expired in 1980; thus, the body mandated to approve such funding requests no long exists.[‡]

Although there has been energetic public debate in Great Britain, Australia, and many other countries, in vitro fertilization in the US is essentially left to the private sector. The lack of public policy and of any public funding means that there is basically no societal regulation, nor any opportunity for public input into the research goals that this society may choose to pursue. The inaction of HEW, now Health and Human Services (HHS), on this matter appears to exhibit a total lack of moral courage. Given the significance of the scientific advances and the importance of the ethical questions they raise, the inaction of this federal agency is inexplicable. On a politically charged topic, however, passivity is an easy solution. It

---

[†]The ethical criteria for research specified the goal of research (to demonstrate the safety and efficacy of clinical in vitro fertilization), required that gamete donors be informed and consenting, set a time beyond which embryos were not to be sustained in the laboratory (14 d), and mandated public notice if the risk of abnormal offspring turned out to be greater than with natural reproduction. In the clinical setting, embryo transfer was to be attempted only if the gametes came from a married couple, and the wife would carry the pregnancy.

[‡]On July 14, 1988, Dr. Otis Bowen (Secretary of HHS) announced the first steps toward formation of a new Ethics Advisory Board. A proposed charter will be published in the Federal Register for comment. If established, the EAB will review applications from researchers applying for federal grants to do fertility research involving early human embryos.

may be politically more expedient to ignore a highly controversial issue than to take and defend a position on it.

Interpretation of the Fundamental Principles

RESPECT FOR PERSONS. The underlying question is: How does the principle of respect for persons apply to early human embryos? The EAB report defined an early human embryo as the conceptus from fertilization to 14 days later. At 14 days, individuation is fixed, differentiation begins with the appearance of the primitive streak, and implantation in the uterus is completed. Because the initial stage of development has the characteristics of cellular life rather than those of an individuated human organism, recent documents have advocated use of the term "preembryo" to designate the undifferentiated, preimplantation stage of human life.[16,17]

The federal regulations on fetal experimentation take no position regarding the moral status of preembryos or what respect for them demands of us. In these regulations, "fetus" refers only to an already-implanted conceptus (45 CFR 46.203).[18] Similarly, legal decisions on abortion deal only with fetal life within a pregnant woman. We have little guidance as to the moral or legal status of preimplantation embryos, or preembryos, in the laboratory.[19-21]

After hearing extensive testimony on the status of preembryos, the EAB concluded that preembryos deserve "profound respect, but...not necessarily...the full legal and moral rights attributed to persons."[22] This conclusion provides more of a negative than a positive directive: the preembryo need not be treated as a full human person. The practical implications are significant, yet not entirely clear. How does one respect an entity, yet not respect it as a person? An interpretation could be so narrow as merely

Ethical Considerations

to indicate that the pre-embryo ought not be counted in the census, or as a dependent for income tax purposes. However, it could be so broad as to suggest that the preembryo be treated like other human tissue in the laboratory, usable for research purposes as long as one handles it with care and only for appropriate purposes (for example, not as the matter of a practical joke).

It is apparent from the EAB report that the members believed that respect for the preembryo should be understood to fall somewhere between the two suggested interpretations. The preembryo could be used for research purposes only under carefully specified conditions[†] and only if each individual research proposal was approved by an Ethics Advisory Board at the federal level.

However, the EAB's 1979 conclusion as to what constitutes sufficient respect for the preembryo was not applied to public policy decisions, and there has been a conspicuous lack of further public debate on the issue in the United States. Whereas most people would probably agree that early embryos should be treated with "profound respect," though not necessarily that accorded to full persons, some people would not concur. A significant minority in our society, and certainly some religious bodies,[23] would hold that the preembryo should be treated as a human person.

Even among those who would not require that we respect preembryos as we respect human persons, there is

---

[†]The ethical criteria for research specified the goal of research (to demonstrate the safety and efficacy of clinical in vitro fertilization), required that gamete donors be informed and consenting, set a time beyond which embryos were not to be sustained in the laboratory (14 d), and mandated public notice if the risk of abnormal offspring turned out to be greater than with natural reproduction. In the clinical setting, embryo transfer was to be attempted only if the gametes came from a married couple, and the wife would carry the pregnancy.

a wide range of disagreement as to what constitutes appropriate respect. Are some types of research disrespectful because their purpose is not significant enough? Are some types of procedures, such as destructive procedures, too disrespectful, whereas observation or karyotyping would be acceptable? Is it disrespectful to intentionally create human embryos for the purpose of research, but permissible to conduct research on "spare" embryos?

These issues are too crucial to be determined solely by scientists in privately operated programs. They are issues for public policy debate, and for public determination of ethical limits and appropriate funding. So far, we have made little progress in interpreting the principle of respect to determine public policy regarding early embryos.

WELL-BEING. The principle of well-being applies to external fertilization in several ways. The first concern is the health and well-being of the fetus and infant that may develop from the laboratory-fertilized embryo. Pierre Soupart and other scientists initially had reservations about the safety of the in vitro process. The EAB supported research to demonstrate and/or improve safety, arguing that it is irresponsible to create new human lives without knowing whether one is, at the same time, causing defects.

A substantial history of in vitro fertilization (10 yr and thousands of births) has now shown that there is not a disproportionate number of birth defects in these infants. However, other problems remain. Preembryos are generally rejected for implantation when they do not "look good," a highly unscientific and subjective judgment. Moreover, the process of cryopreservation or freezing of preembryos is conducted within a context of even less scientific data; the effects of freezing embryos are largely unknown.[24] The ethical principle of well-being requires carefully controlled research to determine the effects of various manipulations on the early embryo, and hence on the health of future children. By its decision not to fund such research, our society permits chance to play a larger than necessary role in relation to birth defects. Concern for well-being mandates that in vitro fertilization be made as safe as possible, unless it is to be totally prohibited.

A second concern is the well-being of the infertile couple who pursue in vitro fertilization. Given the low rate of success (measured in live births) in the majority of in vitro clinics, many couples may be psychologically, economically, and even physically harmed. Improved efficiency would provide significant benefits to the infertile couples who use in vitro fertilization. Controlled research is essential to improve efficiency, and the principle of well-being appears to require such research.[25]

A third, and more hypothetical concern, is the well-being of all future fetuses and infants at risk for genetic diseases and birth defects. If we could prevent many of these problems through embryo research, then society needs to determine whether to support such research. In this situation, the information sought is not directly related to possible harms from the in vitro process; thus, the ethical obligation to conduct such research may diminish. (In general, one is morally required to avoid causing harm, but not necessarily to seek knowledge, even if it could improve human life.) Whether preembryos in the laboratory should be used as a source of information on congenital diseases and their prevention depends on the basic question raised earlier: What is the moral status of preembryos, and what does it mean to respect them?

EQUITY OR JUSTICE. The principle of justice requires that we treat persons fairly, avoid discrimination, and distribute social benefits and burdens in an equitable manner. The EAB could not determine whether public funding of in vitro fertilization research would be a just use of society's scarce resources. Are infertility treatments part of basic medical care? If so, will insurers be required to include them as part of their coverage, and will federal reimbursement programs contain such provisions? If third-party payers do not cover infertility treatments, then are those who are less affluent being denied the means to exercise a "right to reproduce"?[26]

Recently, a new possibility for injustice has arisen because of multiple pregnancies resulting from in vitro fertilizations and other infertility treatments. In a few highly publicized cases, one or more fetuses in a multiple pregnancy have been selectively killed intrauterinally.[27] Although the intention may be to save at least some fetal lives in this high-risk situation, such selective killing appears discriminatory. The fetuses are often selected arbitrarily or because of convenience of location in the uterus. These criteria would not be considered adequate for making life-death decisions regarding born persons; are they fair criteria for determining which fetuses live or die?

Selective abortion could disappear if appropriate limits were accepted by in vitro clinics. Many clinics appropriately and successfully limit the number of oocytes that are fertilized and transferred. Perhaps this sort of limitation should be required, given that implantation of an excessive number of preembryos inevitably leads to either an extremely high-risk pregnancy or the selective killing of fetuses.

Conclusion

Our society has not adequately confronted the question of the moral status of the preembryo. The goals of research involving preembryos may be highly laudatory, but their moral weight can be assessed only in relation to the respect due to the preembryo "subjects" of this research. Until we have a consensus as to what is entailed by respect for preembryos, the principle of well-being lacks a firm foundation in its application to public policy on in vitro fertilization.

Issues raised by the ethical principle of equity are not unique to in vitro fertilization, and these issues call for approaches that are consistent with biomedical decision-making in other areas.

Contract Motherhood

Procedures and Problems

The techniques for assisted reproduction in the second category are undertaken mainly in the clinic (or home), using only the ordinary lab supports of clinical medicine. The ethical problems arise largely from the involvement of a third party, such as a sperm, oocyte, or embryo donor, or a gestational/birthing mother. The practice of artificial insemination by donor sperm (AID) has been used widely for some years, whereas the donation of preembryos for transfer is still a relatively rare practice. The donation of oocytes is also uncommon, except when it is part of a surrogate or contract mother arrangement, in which a woman agrees to be inseminated and to conceive, carry, and give birth to a child that she will give to a contracting couple, the rearing and social parents.

Reproductive arrangements involving the donation of gametes, preembryos, or gestational and birthing services deliberately separate genetic or gestational parenting (or both) from social parenting. Usually, hired professionals facilitate the arrangements for the parties involved. Whether to become involved raises questions—for all participating individuals—about personal and family relationships, the meaning of parenting, the welfare of all concerned, and foresight into future outcomes for society. From society's perspective, what should be the role of public policy with regard to reproductive arrangements transacted privately and with the free consent of everyone involved (except the child to be procreated)?

Public Policy

State statutes on AID were first enacted in the 1950s, and are continually being reviewed and amended. As recently as 1986, states without AID legislation have added such laws to their books.[28]

Presently, contract motherhood poses the greatest challenge for public policy. Its use is extensive and expanding, and it impinges on established aspects of public policy, such as laws on adoption and baby selling.

In Britain, following a recommendation of the Warnock Committee, the Surrogacy Arrangements Act of 1985 was enacted. This Act prohibits, with fines up to 2000 pounds, any agency or professional from negotiating or arranging surrogate motherhood for payment, and from conducting related activities, such as recordkeeping or advertising.[29]

In the well-publicized Baby M case, the New Jersey Supreme Court declared commercial surrogate motherhood contracts illegal,[30] but did not rule on voluntary unpaid surrogacy agreements.[31] The Baby M case spurred the introduction of a plethora of state (and several federal) laws prohibiting or regulating surrogate motherhood contracts. Some are concerned only with commercial surrogacy, whereas others also include voluntary arrangements. Very few of the bills have been passed so far. The complexity of the issues and the multitudinous details to be considered suggest that a substantial period of reflection and debate ought to precede the enactment of statutory legislation.[32]

Interpretation of the Fundamental Principles

RESPECT FOR PERSONS. The principle of respect for persons is violated when persons are exploited. Contract motherhood may be exploitative in a number of ways. For example, a contract mother may be exploited because of her financial or emotional need, or an infertile couple may be financially and emotionally exploited in their desperation to have a child. Professionals may be in a position to exploit the deep human wants and needs of the principal parties to the contract, perhaps for their own financial gain.

Despite the potential for exploitation, is it appropriate for public policy to intervene in order to forestall this possibility? By intervening, society would be responding in a paternalistic way, protecting mature persons from the adverse effects of their own choices. There are precedents for both intervention and nonintervention: we do not allow persons to sell their kidneys or to engage in prostitution, but in most areas we allow freedom in personal choices, even if some persons choose exploitive relationships.

There are persuasive arguments for both positions. With regard to the choices of mature adults, contemporary ethical discussion would support a public policy that allows personal autonomy to operate. The dangers arising from paternalistic social policies are often greater than the risks that individuals assume in making their own choices.

Respect for the children procreated as part of a contractual arrangement raises different issues. Here there is no possibility of obtaining the child's consent. The question is often phrased in a very abstract way: Is this child being used as a means to further the ends of others, rather than valued for himself or herself? It is true that no child chooses his or her own birth or birth circumstances. Furthermore, any child may, in some ways, be used by its parents to satisfy the parents' own needs and life goals. However, many commentators find a significant difference between this situation and that resulting from surrogate motherhood. In such contractual arrangements, the conception of a child is consciously chosen and deliberately constructed as a means to the ends of others, including the contracting couple, the brokers, and the surrogate mother.

In a surrogacy contract, the contracting parties agree on the benefits each shall receive: the contract mother will receive money, the agency and lawyers who facilitate the contract will receive their fees, and the contracting couple will receive an infant. Although such contracts are often described as hiring a woman to perform a service, most contracts require payment of only a small fraction of the total fee (usually 10%) if a live infant does not result from the pregnancy. Thus, these contracts should be regarded as payment for providing a baby to the contracting couple.

Every state has laws against baby selling, based on the belief that to treat a child as a market commodity is to treat him or her as a thing, rather than with the respect that is owed a person. To the extent that surrogacy contracts are a form of baby selling that treat the child as a commodity, the ethical principle of respect for persons would prohibit them.

WELL-BEING. The well-being of the child to be procreated through a surrogacy arrangement, as well as that of the families involved, is at stake. However, American public policy allows individuals freedom in reproductive decisions, even if a decision may not seem conducive to the welfare of the children or the families. Clear proof of child abuse or neglect is required before society deems it appropriate to intervene.

Surrogate motherhood offers society a new method of family building. How should public policy balance society's interest in child protection with its interests in protecting reproductive freedom, in relation to this novel arrangement? To answer this question, we need data that enable us to predict the long-term effects of unconventional procreative arrangements. What can we foresee about the welfare of children born as a result of such arrangements? What are the effects on the contractees and families involved?

It is fair to say that we are not very good at making such predictions. Even in such disastrous situations as the Baby M case, we do not know how baby Melissa may be affected in the future. In his rationale for granting extensive visitation rights to Melissa's birth mother, Judge Birger Sween argued that Melissa "is no less capable than thousands of children of broken marriages who successfully adjust to complex relationships when their parents remarry."[33] However, the complex set of relationships in which Melissa finds herself is a result of arrangements made before her conception; what Judge Sween calls "broken marriages" usually occur after the births of the children involved and for reasons not primarily related to their existence.

The intentional separation of genetic, gestational, and social parenting, a separation intrinsic to contract motherhood, may adversely affect the well-being of the children produced. At the very least, a society that routinely accepts this separation would differ from the one we have today. Questions of parental responsibility for child welfare arise. We have already seen the adverse effects of the failure of some genetic fathers to accept responsibility for the rearing and even the support of children they procreate. In surrogate motherhood, the genetic and gestational mother repudiates such responsibility even before the child is conceived.

Differences in the maternal–fetal bonding that occur in surrogate motherhood may also have implications for the well-being of the child and the surrogate mother. The surrogate mother is asked *not* to bond with the fetal life she carries.[§] Although pregnancy is characterized as a relationship of nurturing, the contract mother is to view herself as a carrier or container.[34] Although there is little psychological literature on sur-rogate pregnancy, there is a great deal of research on ordinary pregnancy and parturition.[35-37] Some authors claim that failure to establish a mother–child relationship after quickening is correlated with psychological maladjustment; not only does a pregnant woman ordinarily enter into a conscious relationship with the fetus at quickening, but in addition, this response is deemed psychologically normal. To ask a woman to deny this relationship is to ask her to do something psychologically unhealthy for herself and perhaps also for the child. Since society has an interest in the prenatal and postnatal development of the next generation, it may want to examine the implications of contract motherhood with regard to these early developmental stages.

Regarding development in later childhood, one could surmise that, as long as surrogate arrangements are novel, the resulting children may suffer by reason of their unconventional procreation. These children may consider themselves odd or fear that others consider them odd.

[§]Note excerpt from a typical contract: "Mary Beth Whitehead understands and agrees that...she will not form or attempt to form a parent–child relationship with any child or children she may conceive, carry to term and give birth to."[38]

Such stigmas would probably lessen if the process became more widely used and accepted, but whatever the degree of societal acceptance of surrogacy, a child's self-esteem may be severely damaged by knowledge of the details of his or her origin. Whereas adopted children may be comforted and supported by the belief that their birth mothers gave them up out of love, some children of surrogate mothers will know that their birth mothers gave them up for money.

EQUITY OR JUSTICE. The principle of equity demands that classes of persons be treated fairly in the distribution of social benefits and burdens. Our society has become very sensitive to institutional arrangements prejudicial to the interests of the groups composing it. The US Supreme Court has prohibited certain inherently unfair practices, for example, segregation of the races in public education. In addition, Civil Rights Acts and other legislation outlaw class-based restrictions on free association, public accommodations, and economic opportunity.

In the practice of contract motherhood, there is a potential for some groups in society to benefit at the expense of others. Currently, it appears that these arrangements benefit mainly the affluent, and burden those who are in a less favorable socioeconomic position. Furthermore, the benefits seem to be predominantly available to men, with the burdens falling on women. The principle of equity requires us to develop policies that minimize inequities of these kinds.

Conclusion

I believe it will be a long time before we are adequately able to predict the effects of surrogacy on the well-being of the children produced. Thus, predictions of future well-being or harm may not offer much guidance for public policy at present. The principle of respect for persons and possibly the principle of equity currently provide more secure bases for the formulation of public policy on surrogacy. These principles point to the need for policies that prohibit surrogate mothers' being paid fees for their "services" or their "products." On the other hand, prohibiting surrogacy on grounds that it has adverse consequences for the well-being of the children produced or the parties involved does not appear warranted at this time.

References

[1]Murray, J. C. (1960) "Should There Be a Law?," Chapter 7 in *We Hold These Truths: Catholic Reflections on the American Proposition* (Sheed and Ward, New York).

[2]National Commission for the Protection of Human Subjects, *The Belmont Report* (Washington, DC: US Government Printing Office, 1978).

[3]President's Commission for the Study of Ethical Problems in Medicine and Biomedical and Behavioral Research, *Summing Up* (Washington, DC: US Government Printing Office, 1983).

[4]Bayles, M. D. (1984) *Reproductive Ethics* (Prentice-Hall, Inc., Englewood Cliffs, NJ).

[5]Overall, C. (1987) *Ethics and Human Reproduction: A Feminist Analysis* (Allen & Unwin, Boston).

[6]Stanworth, M. ed. (1987) *Reproductive Technologies: Gender, Motherhood and Medicine* (University of Minnesota Press, Minneapolis).

[7]Walters, L. (June 1987) Ethics and new reproductive technologies: An international review of committee statements. *Hastings Cent. Rep.* **17(3)**, 3–9 of special supplement.

[8]Council for Science and Society (1984) *Human Procreation: Ethical Aspects of the New Techniques* (Oxford University Press, Oxford).

[9]Bernardin, J. C. (May 28, 1987) Science and the creation of human life. *Origins* **17(2)**, 21–26.

[10]Callahan, S. (April 24, 1987) Lovemaking and babymaking: Ethics & the new reproductive technology. *Commonweal* **114(8)**, 233–239.

[11]Cahill, L. and McCormick, R. (March 28, 1987) The Vatican documents on bioethics: Two responses. *America*, **156(12)**, 246-248.

[12]Walters, W. and Singer, P. (eds.) (1982) *Test-Tube Babies* (Oxford University Press, Melbourne).

[13]Caplan, A. L. (1989) Arguing with success: Is in vitro fertilization research or therapy?, this volume.

[14]Fletcher, J. C. and Ryan, K. J. (1987) Federal regulations for fetal research: A case for reform. *Law, Medicine and Health Care*, **15(3)**, 126–138.

[15]Ethics Advisory Board (June 18, 1979) Protection of human subjects; HEW support of human in vitro fertilization and embryo transfer. *Federal Register*, **44**, 35033–35058.

[16]American Fertility Society (Ethics Committee) (1986) Ethical considerations of the new reproductive technologies. *Fertil. and Steril.* **46(3) Suppl. 1**, 1S–94S.

[17]Voluntary Licensing Authority (Great Britain), Annual Reports 1,2,3 (1986, 1987, 1988). Obtain from VLA Secretary, Medical Research Council, 20 Park Crescent, London W1N 4AL, England.

[18]Fletcher, J. C. and Ryan, K. J.

[19]Bondeson, W. B., Engelhardt, H. T., Jr., Spicker, S. F., and Winship, D. H. (eds.) (1983) *Abortion and the Status of the Fetus* (D. Reidel, Dordrecht, Holland).

[20]Shaw, M. W. and Doudera, A. E. (eds.) (1983) *Defining Human Life: Medical, Legal, and Ethical Implications* (AUPHA Press, Ann Arbor, MI).

[21]Walters, L. B. (ed.) (August 1985) Genetic and reproductive engineering. *J. of Med. and Philos.* **10**, 209–309.

[22]Ethics Advisory Board, 35056.

[23]Brody, B. A. (1989) Current religious perspectives on the new reproductive techniques, this volume.

[24]Ellis, G. B. (1989) Infertility and the role of the federal government, this volume.

[25]Caplan, A. L.

[26]Childress, J. F. (1981) *Priorities in Biomedical Ethics* (Westminster Press, Philadelphia).

[27]Berkowitz, R. L., Lynch, L., Chitkara, U., Wilkins, I. A., Mehalek, K. E., and Alvarez, E. (April 21, 1988) Selective reduction of multifetal pregnancies in the first trimester. *N. Engl. J. of Med.* **318**, 1043–1047.

[28]Andrews, L. B. (October/November 1987) The aftermath of Baby M: Proposed state laws on surrogate motherhood. *Hastings Cent. Rep.* **17(5)**, 31–40.

[29]Brahams, D. (February 1987) The hasty British ban on commercial surrogacy. *Hastings Cent. Rep.* **17**(1), 16–19.

[30]Merrick, J. (1989) The case of Baby M, this volume.

[31]Hanley, R. (February 4, 1988) Surrogate deals for mothers held illegal in Jersey. *New York Times*, pp. 1, 14.

[32]Andrews, L. B.

[33]Hanley, R. (April 7, 1988) Surrogate mother wins visitation rights. *St. Paul Pioneer Press Dispatch*, pp. 1A, 5A.

[34]Rothman, B. K. (1989) Recreating motherhood: Ideology and technology in contemporary society, this volume.

[35]Bibring, G., Dwyer, T. F., Huntington, D. S., and Valenstein, A. F. (1961) A study of the psychological processes in pregnancy and of the earliest mother–child relationship. *Psychoanal. Study Child* **16**, 9–72.

[36]Jessner, L. (1966) On becoming a mother, in *Conditio Humana*, (von Baeyer, W. and Griffith, R. M., eds.), Springer-Verlag, Berlin, 102–114.

[37]Jessner, L., Weigert, E., and Foy, J. L. (1970) The development of parental attitudes during pregnancy, in *Parenthood: Its Psychology and Psychopathology* (Anthony, E. J. and Benedek, T., eds.), Little, Brown and Co., Boston, 209–244.

[38]Arditti, R. (1988) A summary of some recent developments on surrogacy in the United States. *Reproductive and Genetic Engineering* **1**(1), 51–64.

Treatment of Infertility
and
Assisted Reproduction

Medical Techniques for Assisted Reproduction

George Tagatz

A couple having intercourse during the periovulatory period has no more than a one in five chance of conceiving a pregnancy that will result in the delivery of a mature infant. Since fertilization occurs in at least four of five cycles, this is an apparently poor rate of success. However, with regard to early reproductive loss, failure is more appropriately viewed as a natural fail-safe mechanism to eliminate abnormal gametes and embryos. Spontaneous abortions serve as the process for the selection of normal and near-normal embryos and fetuses. Pregnancies are aborted with decreasing frequency as pregnancy progresses; approximately 70% of fertilized oocytes will be rejected before a pregnancy is recognized by the usual clinical criteria. After the first trimester of pregnancy, the majority of abnormal pregnancies have been eliminated; pregnancy loss after 14 weeks of gestation is negligible.

Utilizing existing data, one can construct a model of the expected fates of 100 mature oocytes released singly from the ovaries at midcycle. Judging from our experience with in vitro fertilization, 15% of the mature oocytes released at ovulation will not be fertilized. From the time of fertilization on day 14 to presumed implantation on day 21 of an idealized 28-day cycle, another 25% of oocyte-progeny are rejected in the preembryo stage. From the time of implantation to the clinical recognition of a pregnancy at 3½–4 weeks after ovulation (5½–6 wks from the last menstrual period), an additional 35% of the oocyte-progeny will be discarded during early embryonic development.[1] Although 15–20% of clinical pregnancies will be aborted after the usual signs and symptoms of pregnancy become apparent, this loss comprises only four to five of the original hypothetical 100 oocytes (oocyte-progeny).[2] In summary, of 100 oocytes, only 24 are expected to be fertilized and develop to the stage where the mother has a clinically recognized pregnancy, and only 20 will proceed to the delivery of a mature infant (see Table 1).

Pregnancy loss is attributable to either genetic and developmental defects of the gametes and preembryos, or to the maternal environment.[3] The intricacies of embryogenesis culminating in the development of a highly complex, normal human being are incompletely defined. Methods currently used to analyze aborted tissues can detect only the most obvious abnormalities in the rejected embryos. Since chromosomal and anatomic defects are discernible in nearly half of the fetoplacental tissue samples aborted after clinical pregnancy is recognized, it is assumed that intrinsic defects of the germ plasm are present in the majority of instances of earlier pregnancy loss. This conjecture is also supported by the relative infrequency of pregnancy loss that can be attributed to abnormalities of the maternal environment. Thus, early pregnancy loss is viewed as a salutary and efficient process of selection of embryos destined to develop into healthy infants. Of those fetuses that develop to maturity, only 3–5% will have serious congenital abnormalities.

Table 1

In Couples with Normal Fertility Potential, the Fates of 100 Mature Oocytes Released Singly from the Ovaries at Midcycle Are Approximated

Stage of development	Surviving oocytes and oocyte progeny	Event or diagnostic criteria
Oocytes	100	Fertilization
Zygotes	85	
Preclinical pregnancies	60	Implantation Positive biochemical assay (βHCG)
Clinical pregnancies	24	Delayed or missed menses
		Ultrasound discerns 6 weeks from last menses, 4 weeks after ovulation
Mature gestation	20	Infant delivered at ±40 weeks of gestation

During trials utilizing relatively crude techniques for the extracorporeal manipulation of oocytes, spermatozoa, and preembryos, it was presumed that the same process of maternal rejection would ensure a successful outcome. Experience has confirmed that abnormalities in the liveborn offspring resulting from in vitro fertilization or from insemination with cryopreserved semen do not exceed the rate of congenital anomalies in liveborn infants conceived by sexual intercourse. Abnormal gametes and embryos, however derived, are efficiently discarded during the early stages of human gestation.

From a biological perspective, couples with normal fertility potential have, during each cycle of exposure, only a one in five chance of conceiving a pregnancy that will result in a mature offspring. The pool of couples attempting pregnancy is decreased by the number who have achieved pregnancy during previous cycles; thus, a human fecundity rate of 20% per cycle does not result in 100% of couples achieving pregnancy within five cycles of exposure. In the general population, 80% of couples conceive a clinically recognizable pregnancy during the first 12 months of exposure, and 90% are successful by the end of 24 months or two years. These demographic statistics define the infertile population as consisting of the 10% of couples who are unable to conceive a clinical pregnancy before two years have elapsed. Some of these couples will be aided by medical intervention; others will not. Spontaneous cure continues to be a function of time among "infertile" couples who are ultimately found to have no apparent cause for infertility; in this group, the spontaneous pregnancy or cure rate is about 4% per cycle.[4]

Among the couples who have recognizable reproductive defects, infertility may result from absolute barriers to conception, such as completely obstructed fallopian tubes or complete absence of spermatozoa. In most instances, however, relative causes are responsible for inability to conceive a clinical pregnancy; by analogy these would include pelvic adhesions partially preventing access from the fallopian tube to the ovarian site of ovulation, or less than an optimal concentration of spermatozoa.

The causes of infertility are determined by assessing the anatomical features and physiological functions required for fertilization and early pregnancy maintenance. The large number of potential defects can be ascertained by dissecting the components of four theoretical compartments (see Table 2).

1. Semen quality and ejaculation: Successful function requires normal spermatogenesis and hormone secretion; the intact male reproductive tract should deliver an ejaculate containing at least 20 million motile spermatozoa.

2. Couple factors: Histochemical and secretory changes in the female reproductive tract are optimal for survival and migration of spermatozoa at the time of ovulation, which occurs at the midpoint of the menstrual cycle. Sperm must be deposited and survive in the cervical mucus at midcycle, prior to migrating through the uterus to the site of fertilization at the end of the fallopian tube.

3. Anatomic integrity of the female reproductive tract: Fertilization requires a clear channel for upward migration of spermatozoa and access of the sperm to the oocyte released from the ovary. Successful implantation requires downward movement of the preembryo from the fallopian tube to a favorable site in the uterus.

4. Physiological integrity of the female reproductive tract: A normal, ovulatory menstrual cycle requires that the ovary integrate hormone secretion with the maturation of the oocyte and follicle. In a dynamic feedback re-

Table 2
Evaluation of the Infertile Couple

Female	Male
Ovulation	Semen analysis
Hormone assays	
Basal body temperatures	
Progesterone assays	
Endometrial biopsy	
Follicle development	
Ultrasound	
Pelvic anatomy and physiology	
Palpation (pelvic exam)	
Hysterosalpingogram	
Laparoscopy; hysteroscopy	

Couple

Intercourse history
Sims-Huhner postcoital test

lationship controlled by the circulating levels of ovarian hormones, the ovary both responds to and modifies the gonadotropin signals originating in the pituitary. Pituitary responsiveness to ovarian feedback requires the additional input of hourly pulses of GnRH, a decapeptide secreted by the hypothalamus. The GnRH secretion is, in turn, responsive to multiple neurotransmitters secreted at the hypothalamic terminals of nerve cells originating in the brain and brain stem. In addition to the self-regulating activity of ovarian hormones on pituitary signals, the estrogen and progesterone provide the sequentially graded stimuli that prepare the reproductive tract for sperm reception, migration, fertilization, implantation, and early pregnancy maintenance.

The enormously complex and delicate interrelationships required for normal reproductive function are readily disrupted by both local abnormalities of the reproductive tracts and systemic disease in either partner. Potential etiologies include hereditary and acquired disorders, which may be genetic, infectious, immunologic, hormonal, metabolic, neoplastic, or neuropsychiatric in origin.

Couples with preclinical or unrecognized early pregnancy loss comprise a second group of apparently infertile partners. The group is defined by sensitive assays that detect the presence and subsequent disappearance of specific pregnancy hormones in maternal serum after presumed implantation, but prior to the appearance of the signs of clinical pregnancy. In the majority of instances, the loss results from a primary defect in the developing preembryos. Although probably of lesser frequency, a maternal illness or disease may produce a defective environment for the normally developing preembryo, including: inadequate hormonal support of the endometrium; immunological rejection; and viral, bacterial, and protozoal infection.

The previous considerations provide a useful background for assessing the results of various treatment regimens. The complexity of the reproductive process presents multiple discernible sites for potential malfunction during a given cycle. Probably an equal or greater number of critical processes are not even identified by the clinical studies of infertility. Correction of a recognized solitary defect in an infertile couple will result in a fecundity rate not exceeding that observed in the normal population. A clinical pregnancy rate of about 60% within six cycles of treatment is a reasonable, if not optimistic, expectation.

An excellent example of a corrected solitary defect is surgical reanastomosis of normal fallopian tubes for the purpose of reversing a previous sterilization procedure. However, since most of these women have previously demonstrated normal fertility potential, the expected results are skewed toward a higher pregnancy rate. Laparoscopic resection of minimal peritubal adhesions would also seem to correct an isolated defect in the female partner of an infertile couple. In this situation, unrecognized microscopic tubal disease, which frequently coexists with the clinically observable scar tissue around the tubes, will reduce the likelihood of a successful outcome in an unpredictable fashion.

The results of treatment with pituitary gonadotropins provide some insight into the problems presented by multiple reproductive defects in both partners. During the 1970s, pituitary gonadotropins were initially utilized to stimulate ovulation. Treatment was reserved for women who were selected for a solitary defect of ovulation, and their partners were required to have semen of normal quality. Within this group of carefully selected couples, we achieved pregnancy rates with gonadotropin therapy of about 25% per cycle.[5] Currently, gonadotropins are administered to achieve multiple ovulations/cycle in the female partners (who may already ovulate) of couples with coexistent causes for infertility, such as oligospermia. When the reasons for using gonadotropins are different and the couple has coexistent defects of reproduction, the results of gonadotropin therapy in the expanded treatment group are obviously not comparable to the results of gonadotropin therapy in the highly selected group. In the expanded group with multiple causes for infertility, our pregnancy rate is 10% per cycle. The heterogeneity of the treated groups and the indications for treatment constitute bias of selection or ascertainment that often invalidates comparisons of results obtained with a single treatment modality. Most importantly, one can anticipate only a limited degree of success in the treatment of infertile couples with multiple reproductive abnormalities.

Gonadotropin therapy also introduces the topic of risks in infertility therapy, since major problems are associated with this treatment regimen. An increase in multiple gestations is a predictable side effect of simultaneous ovulation from several sites on the ovaries. With strict indications for the use of gonadotropins, the occurrence of triplet and higher-multiple pregnancies was so rare as to continue to be viewed as a curiously delightful outcome for couples seeking families. However, with the recent expansion of indications for the use of gonadotropins, including in vitro fertilization, the accumulated experience indicates that high-multiple pregnancies constitute a serious medical complication. Adverse outcomes for the mothers and offspring are so common and severe as to prompt some physicians to advise selective abortion of fetuses in excess of two. Preliminary clinical experience indicates the procedure is reasonably effective. The issue of selective abortion or embryo reduction is subject to the same ethical and legal considerations that apply to all elective abortions.[6-8]

Concomitant with the recruitment and maturation of multiple follicles, the overall dimensions of the ovaries increase significantly. Ten percent of stimulation cycles result in excessive enlargement of the ovaries. In about 1% of cycles, ovarian enlargement is associated with a leakage of fluid into the cavities of the abdomen and chest that leads to a decrease in the volume of fluid retained in blood vessels. This hyperstimulation syndrome has terminated in kidney failure and death for several women throughout the world. Although the hyperstimulation syndrome is not totally preventable, current regimens for gonadotropin administration incorporate blood assays of estradiol and ultrasound examinations of the ovarian follicles, in an attempt to control the degree of stimulation.

Traditional therapy of infertility has focused upon correction of defects in the prospective social parents. Accordingly, the hereditary, gestational, and social relationships of parents to offspring have always been considered to be identical. In the 1973 *Roe v. Wade* decision, the Supreme Court of the United States effectively abolished legal prohibitions concerning a woman's freedom to have an elective abortion during the first trimester of pregnancy. This has resulted in an increasing number of elective abortions and a consequent reduction in the number of children available for adoption. The virtual elimination of this traditional and ethical solution has increased the desire of infertile couples to pursue alternative medical regimens. Among these are techniques for substitution therapy with donor spermatozoa.

The effectiveness of achieving pregnancies with donor insemination for couples with male infertility is well documented. In a program using fresh donor semen, we achieved pregnancies for nearly 500 couples from 1973 to 1985.[9] During the era in which we used fresh donor semen, issues of safety concerned the transmission of hereditary defects and recognized infectious agents. Screening procedures limited insemination-associated infections to rare occurrences and resulted in the conception of infants with fewer congenital abnormalities than are found in the general population. Only four of the nearly 500 babies had either a chromosomal defect (21-trisomy at birth; 16-trisomy electively aborted after genetic amniocentesis revealed the defect) or a congenital anomaly recognized at birth (club foot; cutaneous hemangioma of the chest).

The epidemic of the acquired immune deficiency syndrome (AIDS) and the documented transmission of the causal human immunodeficiency virus (HIV) in fresh and frozen donor semen introduced a uniquely devastating complication to the risk/benefit equation for insemination with fresh donor semen.[10] In 1987, it was estimated that over one million Americans were infected with the HIV virus.[11] After an incubation period of variable duration, many infected individuals will develop AIDS or the AIDS-related complex (ARC). Preventive measures are not reliable, and the disease is resistant to treatment. More than 90% of individuals with AIDS will die within three years of contracting the disease.

Although the HIV virus has been transmitted in frozen as well as fresh semen, cryopreservation provides a method for storage or quarantine of semen specimens during the many weeks from the time of donor exposure until his HIV tests are expected to indicate infection. Depending upon the tests used, the time interval is at least 100 days. The donor is tested for infection with the AIDS virus no sooner than 100 days after the last of the pooled specimens are stored. If the donor tests are negative, the specimens are cleared for use; if the tests are positive, the entire pool of specimens is discarded. Unfortunately, although freezing provides the means to maximize safety, cryopreserved semen has been significantly less effective than fresh semen in achieving pregnancies; in our local community, previous experience with frozen semen obtained from commercial banks has been disappointing.

The emergence of this almost uniformly fatal illness transmissible by fresh semen, together with the existence of a poorly effective but potentially safe alternative, provided an appropriate focus for reassessing our responsibilities in the treatment of infertility. Patients with infertility are usually healthy individuals who seek medical assistance in having children. When assessed objectively, the goal of reproduction justifies only minimal procedural risks, since the proposed benefit does not involve the customary considerations of saving life or restoring health. Yet each person's desire to perpetuate their unique hereditary characteristics may be a compelling motive, resistant to limitations imposed by concerns for personal risk. We believe that it is the responsibility of the community of physicians to weigh the risks and benefits of proposed therapy carefully and to protect vulnerable patients from undertaking procedures with inordinately high risk. The risk of death from an infectious disease has an aura of unreality to an antibiotic generation that has developed an overriding faith in medical miracles. Since my personal experience as a physician did not include a uniformly fatal infectious disease, I assumed that our patients would be more likely to assess the risk/benefit ratio with regard to the rare likelihood of transmission of the HIV virus, rather than the devastating outcome if they contracted AIDS.

We considered the ethical dilemma involved in changing from fresh to cryopreserved semen. The theoretical likelihood of increasing safety at the expense of an established reduction in efficacy was problematic.[12-14] We resolved the problem by declaring a moratorium on our fresh semen donor program in October 1985. In February, 1988, the American Fertility Society (AFS) officially recommended the abolition of donor programs utilizing fresh semen.[15] The AFS also proposed criteria for quarantine and screening of donors and semen to prevent the transmission of the HIV virus in programs utilizing cryopreserved spermatozoa.[16]

The redevelopment of our donor program has entailed the separate tasks of projecting effectiveness and provid-

Table 3

Evaluation of Semen Donors

Comprehensive medical and social history of donor and family (American Fertility Society Protocol)
Fertility potential of semen
 Semen analysis, before and after cryopreservation.
 Hamster egg penetration assay after 1 week of cryopreservation
Laboratory evaluation of semen specimens and donors

---

ing safety. To maximize effectiveness, we require the recovery of at least 40 million motile spermatozoa in each thawed specimen and utilize only donors with spermatozoa that, after freezing and thawing, are capable of penetrating a zona-free hamster egg.[17] To maximize safety, we have collaborated with experts in infectious disease to develop screening procedures that will detect those donors who are HIV carriers before their cryopreserved spermatozoa can be used for insemination. The enormity of the project can be appreciated from Tables 3 and 4.

With the birth of Louise Brown in 1979, extracorporeal fertilization and embryo transfer became a reality, and Steptoe (Patrick) and Edwards (Robert G.) became household words. In the procedure for in vitro fertilization (IVF), fertilization occurs within a laboratory incubator about 24 hours after retrieved oocytes and capacitated spermatozoa are commingled in culture medium. The majority of the fertilized oocytes (zygotes) will begin cleaving, or dividing into preembryos, within the next 24 hours. At this time, about 48 hours after oocyte retrieval, the newly formed embryos are transferred from the incubation dish to the uterus of the gestational mother. The initial application of IVF was to treat infertile women with absent or

Table 4
Laboratory Evaluation
of Semen Specimens and Donors

	Initial *	Final **	Final +100 days***
Blood type (ABO, Rh) and antibody screen	X		
Blood leukocyte karyotype		X	
Serologies:			
RPR (VDRL)	X		X
HTLV3 (HIV)			
Elisa	X		X
Western blot	X		X
Herpesvirus (HSV-2; HSV-1)	X		X
Hepatitis A (Ab)			X
Hepatitis B (Ag, Ab)			X
Non-A, non-B hepatitis			
Bilirubin			X
GGT			X
ALT (SGPT)			X
Cytomegalovirus	X		
Peripheral blood leukocyte culture			
HIV (HTLV3)		X	
Semen and urethral cultures			
Gonoccus	X	X	
Chlamydia (urethra)	X	X	
Ureaplasma (urethra)	X	X	

(continued)

Table 4 (*continued*)

	Initial*	Final**	Final +100 days***
Listeria monocytogenes (cold enrichment)	X	X	
Routine aerobic	X	X	
Specific donor groups			
Jewish			
Tay Sachs	X		
Black (HgbS, C)			
Hemoglobin electrophoresis	X		
Mediterranean & East Asian (Thalassemias)			
Hemoglobin electrophoresis	X		

*Initial screening is performed prior to obtaining specimens for cryopreservation.

**Final screening—repeat cultures are obtained from the last semen specimen submitted for cryopreservation; peripheral blood karyotype is initiated.

***Final + 100 days—serologies are obtained 100 days after the last semen specimen has been submitted.

damaged fallopian tubes. Although the procedure of gamete intrafallopian transfer (GIFT) also requires the retrieval of oocytes, GIFT differs significantly from IVF, in that fertilization occurs in the gestational mother (in vivo), and the gamete recipient must have at least one normal fallopian tube. In GIFT, retrieved oocytes are loaded into the same catheter with capacitated spermatozoa; oocytes and sperm may be separated in the catheter by a bubble of air. Within minutes after retrieval of the oocytes, the loaded catheter is inserted into the fallopian tube(s) of the gestational mother, where the sperm and oocytes are deposited, and fertilization occurs subsequently. The resultant preembryos must traverse the fallopian tube to be implanted in the uterus. A representative experience with IVF-ET (embryo transfer) in the successful program at the University of Minnesota is presented in Table 5. The nationwide experience is reported in ref. 18.

In vitro fertilization makes it possible to transfer the preembryo derived from the gametes of genetic parents to a separate gestational mother. As for other infants, the newborn could be adopted and reared by social parents with no relationship to the genetic or gestational parents.

Table 6 outlines many of the alternatives currently available for the treatment of infertility. Part I illustrates the multiple possibilities for using husband, wife, or donors as sources of spermatozoa and oocytes for the various reproductive procedures, such as IVF and AID. In Part II, the procedure of lavaging the uterine cavity to recover the preembryo before implantation is presented as a further startling, but successful possibility for intervention in the reproductive process. Finally, Part II B recognizes that for years after initial storage cryopreserved preembryos can be transferred back to the donor or to an alternative recipient.

Because of the legal and social implications resulting from current conventions for contracts utilizing surrogate mothers, a separate discussion is warranted. A surrogate mother is inseminated with sperm obtained from the husband of the couple who desire to adopt the child. Although apparently analogous to AID, surrogate mothering differs in several important respects. As currently implemented, the surrogate mother knows the identity of the biological

Table 5
In Vitro Fertilization and Embryo Transfer
(University of Minnesota)

	Phase I, 8/83–6/85	Phase II, 7/85–12/86	Phase III, 1/87–12/87
Laparoscopic retrievals	32	56	52
Ova retrieved	30 (94%)*	49 (88%)	51 (98%)
Ova fertilized	24 (75%)	45 (80%)	44 (85%)
Preembryo transfers	22 (69%)	37 (66%)	43 (83%)
Clinical pregnancies (ultrasound discerns fetus)	0	13 (23%)	11 (20%)
Births and ongoing pregnancies	0	8 (14%)	8 (15%)
Abortions	0	4	2
Tubal ectopic pregnancies	0	1	1

*Numbers in parentheses are the percentages of total laparoscopic retrievals in each phase.

father and his spouse (the prospective mother of rearing), resulting in an arrangement that forfeits the critical factor of anonymity. Only the surrogate mother is exposed to the medical risks of pregnancy. She alone experiences the psychological bonding of pregnancy, although she may share the experience of gestation and delivery with the prospective parents. With placement of the child for adoption, she may experience the anxiety and grief associated with separation from her offspring. Were the donor and surrogate to remain anonymous, as is the practice in most

Table 6
Reproductive Techniques
for Fertilization and Embryo Transfer

I. Fertilization: by source of gametes (sperm and oocyte) and site of fertilization (uterus or test tube)

Sperm		Oocyte	
Donor	Husband	Wife	Donor
In Vivo Fertilization (Uterus)			
	Coitus	Coitus	
	AIH[1]	AIH	
AID[2]		AID	
	AID surrogate[3]		AID surrogate
	GIFT	GIFT	
GIFT[4]		GIFT	
	GIFT		GIFT surrogate
(GIFT)	Preembryo donors		(GIFT)
In Vitro Fertilization (Test Tube)			
	IVF	IVF	
IVF[5]		IVF	
	IVF		IVF
(IVF)	Preembryo Donors		(IVF)

II. Postfertilization disposition of preembryos for gestation

   A. *Immediate,* by site of fertilization

In vivo (uterus)	In vitro (test tube)
Retain in uterus of fertilized woman	Transfer to uterus of oocyte donor OR hormonally syn chronized woman

(continued)

Table 6 *(continued)*

Retrieve from uterus
    of fertilized woman;
    transfer to second
    uterus of hormonally
    synchronized woman

B. *Delayed,* after cryopreservation

[1]AIH: homologous (husband) insemination.
[2]AID: artificial insemination with donor semen.
[3]Surrogate: Woman other than wife intended to
    carry pregnancy or serve as preembryo donor.
[4]GIFT: gamete intrafallopian transfer.
[5]IVF: In vitro fertilization.

donor semen programs, the failure of the surrogate to surrender the child to an intermediary would simply void an individual contract. The surrogate could become the rearing social parent as well as the genetic and gestational mother, and the intermediary could subsequently engage another surrogate for the semen donor and his wife.

Other authors in this book consider the ethical impact of reproductive techniques that offer seemingly limitless opportunities for departure from the traditionally unified biological and social components of parenting. Although they may not be obvious in discussions relating to specific issues, one can discern several themes or distinct philosophical positions regarding reproductive technologies:

1. Reproduction is acceptable only when it maintains the "natural biological integrity of the sexual act." Restated: Morally acceptable procreation requires sexual intercourse within marriage.

2. Socially responsible permissiveness. The means and ends of reproductive technologies are morally accept-

able if they ensure the dignity and rights of the active participants (parents, donors, surrogates, and offspring), and conform to societal mores and laws regarding marriage, children, and families.

3. Absolute permissiveness or technological imperative. If the means are available to achieve desired ends, the technology should be provided, regardless of the consequences for the participants and society. Restated: What can be done, should be done. The very statement of this view suggests that scientists constitute a coalition of programmed, impersonal, and insensitive zealots that threatens the humanistic traditions of society. To the contrary, the scientific community is composed of a group of individuals with special technical skills who, nonetheless, represent the full spectrum of societal, humanistic, and ethical views.[19]

The involvement of the physician in the implementation of the new reproductive techniques is initiated by the desire of infertile couples, conditioned by the physician's ethical perspectives, and regulated by the legal limitations imposed by society. Having accepted the challenge, it is the sole responsibility of the physician to provide standards of care that incorporate safety and effectiveness as coequal variables in the formula for success.

Acknowledgment

The author gratefully acknowledges the editorial assistance of Ms. Shelley Mortenson.

References

[1]Edmonds, D. K., Lindsay, K. S., and Miller, J. F. ,Williamson, E., and Wood, P. J. (1982) Early embryonic mortality in women. *Fertil. and Steril.* **38**, 447–453.

[2]Roth, D. B. (1963) The frequency of spontaneous abortion. *Int. J. Fertil.* **8,** 431–434.

[3]Warburton, D. (1987) Reproductive loss: How much is preventable? *N. Engl. J. Med.* **316,** 158–160.

[4]Wood C., Baker G., and Trounson A. (1984) Current status and future prospects, in *Clinical In Vitro Fertilization* (Wood, C. and Trounson A., eds.), Springer-Verlag, New York.

[5]Notation, A. D., Tagatz, G. E., and Steffes, M. W. (1978) Serum 17 beta-estradiol: Index of follicular maturation during gonadotropin therapy. *Obstet. Gynecol.* **51,** 204–209.

[6]Evans, M.I., Fletcher, J. C., Zador, I. E., Newton, B. W., Quigg, M. H., and Struyk, C. D. (1988) Selective first-trimester termina- tion in octuplet and quadruplet pregnancies: Clinical and ethical issues, *Obstet. Gynecol.* **71,** 289–296.

[7]Berkowitz, R. L., Lynch, L., Chitkara, V., Wilkins, I. A., Mehalek, K. E., and Alvarez, E. (1988) Selective reduction of multifetal pregnancies in the first trimester. *N. Engl. J. Med.* **318,** 1043–1047.

[8]Hobbins, J. C. (1988) Selective reduction—A perinatal necessity? *N. Engl. J. Med.* **318,** 1062, 1063.

[9]Tagatz, G. E., Gibson, M., Schiller, P., and Nagel, T. C. (1980) Artificial insemination utilizing (fresh) donor semen. *Minn. Med.* **63,** 539–541.

[10]Stewart, G. J., Cunningham, A. L., Driscoll, G. L., Tyler, J. P. P., Barr, J. A., Gold, J., and Lamont, B. J. (1985) Transmission of human T-cell lymphotropic virus type III (HTLV-III) by artificial insemination by donor. *Lancet* **2,** 581–584.

[11]Ho, D. D., Pomerantz, R. J., and Kaplan, J. C. (1987) Pathogenesis of infection with human immunodeficiency virus. *N. Engl. J. Med.* **317,** 278–286.

[12]Iddenden, D. A., Sallam, H. N., and Collins, W. P. (1985) A prospective randomized study comparing fresh semen and cryopreserved semen for artificial insemination by donor. *Int. J. Fertil.* **30,** 54–56.

[13]Smith, K. D., Rodriquez-Rigau, L. J., and Steinberger, E. (1981) The influence of ovulatory dysfunction and timing of insemination on the success of AID with fresh or cryopreserved semen. *Fertil. Steril.* **36,** 496–502.

[14]Richter, M. A., Haning, R. V., and Shapiro, S. S. (1984) Artificial donor insemination: Fresh versus frozen semen: The patient as her own control. *Fertil. Steril.* **41,** 277–280.

[15] Peterson, E. P., Alexander, N. J., and Moghissi, K. S. (1988) A.I.D. and AIDS—Too close for comfort. *Fertil. Steril.* **49,** 209, 210.

[16] The American Fertility Society (1988) Revised new guidelines for the use of semen-donor insemination. *Fertil. Steril.* **49,** 211.

[17] Bordson, B. C., Ricci, E., Dickey, R. P., Dunaway, H., Taylor, S. N., and Curole, D. N. (1986) Comparison of fecundability with fresh and frozen semen in the therapeutic donor insemination. *Fertil. Steril.* **46,** 466–469.

[18] The American Fertility Society Special Interest Group (1988) In vitro fertilization/embryo transfer in the United States: 1985 and 1986 results from the National IVF/ET Registry. *Fertil. Steril.* **49,** 212–215.

[19] Callahan, S. (1987) Lovemaking and babymaking. *Commonweal* **114,** 223–229.

Infertility and the Role of the Federal Government

Gary B. Ellis

Infertility: Problems, Treatments, Costs

Most couples require no assistance to have children. Some couples do not want to have children. This chapter is about the estimated 2–3 million American couples who want to have a baby, but who either need medical help to do so or will remain frustrated in their desire.

Infertility has always been a concern for those affected, but beginning with the birth in 1978 of baby Louise Brown—the first person conceived outside of a human body—and continuing through the 1986–88 surrogate-mother case of Baby M, the public's collective imagination has been captured by new options in the age-old process of procreation.

In mid-1988, the Office of Technology Assessment (OTA), a nonpartisan analytical agency that serves the US Congress, issued the results of its 23-month study, *Infertility: Medical and Social Choices*.[1] The 402-page report analyzes the scientific, economic, legal, and ethical considerations involved in both conventional and novel reproductive technologies. This chapter reviews the report's discussion of the problems of infertility, options available, and several recommendations for US government policy.

Statistics on Infertility

Infertility, generally defined as the inability of a couple to conceive after 12 months of intercourse without contraception, affects an estimated 2.4 million married couples (data from 1982) and an unknown number of would-be parents among unmarried couples and singles. The sole reliable source of demographic information about infertility in the United States is a series of national surveys conducted by the National Center for Health Statistics. The most recent was conducted in 1982;[2] a new survey of about 10,000 women was carried out between January and June 1988, and data will be available later in 1990. In 1982, an estimated 8.5% of married couples with wives aged 15–44 were infertile.

The overall incidence of infertility remained relatively unchanged between 1965 and 1982. Childlessness, or primary infertility, has increased and affects about one million couples. Secondary infertility (in which couples have at least one genetic child) has decreased and affects about 1.4 million couples.

Causes of Infertility

Among women, three factors often contribute to infertility: problems in ovulation; blocked or scarred fallopian tubes; and endometriosis (the presence in the lower abdomen of tissue from the uterine lining). Infections with sexually transmitted diseases—principally chlamydia and gonorrhea—are an important cause of damaged fallopian tubes. Among men, most cases of infertility are a consequence of abnormal or too few sperm. For as many as one in five infertile couples, a cause is never found.

Factors contributing to infertility include the age of the prospective mother. The probability of infertility increases somewhat after age 30 and significantly more after age 35. Although no one social prescription fits all couples in all circumstances seeking to conceive, biology dictates that, to maximize the chance of a natural conception, a couple should maximize the number of months or years devoted to attempting it. A woman's reproductive lifespan is circumscribed, and whenever the decision to procreate is made, the chance of success generally depends on the number of months during which conception is attempted. The probability of conception is reduced both by delaying childbearing and by condensing attempts into a relatively short time period.

Treatment of Infertility

Not all infertile couples seek treatment. An estimated 51% of couples with primary infertility and 22% with secondary infertility seek treatment.

Although there has been no increase in either the number of infertile couples or the overall incidence of infertility in the population, the number of office visits to physicians for infertility services rose from about 600,000 in 1968 to about 1.6 million in 1984. Concomitant increases occurred in the memberships of the American Fertility Society and other professional organizations for physicians who treat infertile patients (see Fig. 1).

There are at least five important factors leading to an increase in requests for infertility services, even absent any increase in its incidence (see Table 1):

- More couples with primary infertility
- Increasing proportion of infertile couples seeking care

Married Couples and Infertility, 1965–82

Infertile couples, 1965–82 (percent of married couples with female age 15 to 44)

(11.2%), (10.3%), (8.5%)

Physician Office Visits for Infertility, 1966–84

American Fertility Society Membership, 1965–86

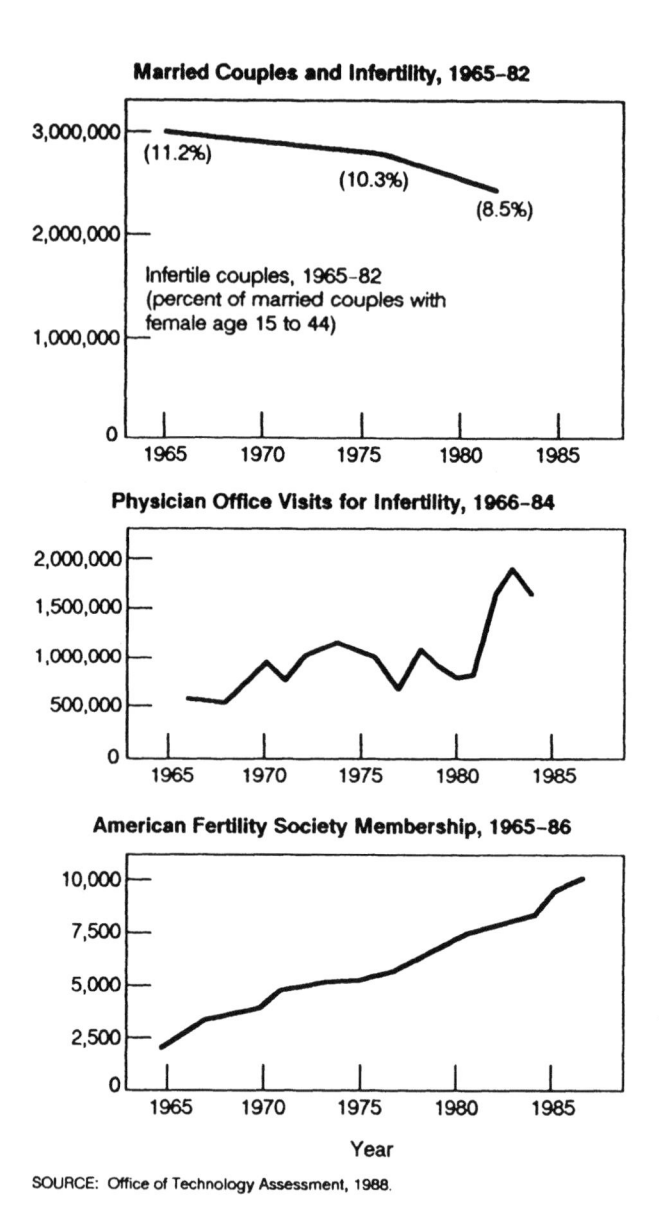

Year

SOURCE: Office of Technology Assessment, 1988.

Fig. 1. Trends related to infertility.

Table 1*
Some Causes of Increasing Requests for Infertility Services in the 1980s

More couples with primary infertility	Increasing proportion of infertile couples seeking care	Increasing number of physicians providing infertility services	More conducive social milieu	Evolution of new reproductive technologies
Aging of the baby-boom generation	Decreased supply of infants available for adoption	Greater demand from private patients	Baby-boom generation expects to control their own fertility	Artificial insemination
Delayed childbearing; more people in higher risk age groups	Heightened expectations	More sophisticated diagnosis and treatment	Profamily movement	Surrogate motherhood
Childbearing condensed into shorter intervals	Larger number of people in higher income brackets with infertility problems	About 180 sites in the US offering in vitro fertilization or gamete intrafallopian transfer	Increased discussion of sexual matters because of the AIDS epidemic	In vitro fertilization (IVF)
Delayed conception resulting from prior use of oral contraceptives	Larger percentage of infertile couples are primarily infertile		Extensive media coverage	Gamete intrafallopian transfer (GIFT)
				Cryopreservation

*Adapted from S.O. Aral and W. Cates, Jr. (1983) The Increasing Concern with Infertility: Why Now? *JAMA* 250, 2327–2331.

- Increasing number of physicians providing infertility services
- More conducive social milieu and
- Evolution of new reproductive technologies.

Among infertile couples seeking treatment, 85–90% are treated with conventional medical and surgical therapy. Medical treatment ranges from instructing the couple in the relatively simple methods of pinpointing ovulation to more complex treatments involving ovulation induction with fertility drugs and artificial insemination. Surgical treatments also span a wide range of complexity, from ligation of testicular veins for eliminating varicocele to delicate microsurgical repair of reproductive tract structures in both men and women. Beyond being physically invasive, treatment is often emotionally taxing. Ovulation induction, surgery, and artificial insemination are still the most widely used and successful approaches to overcoming infertility.

Two new, noncoital reproductive technologies—in vitro fertilization (IVF) and gamete intrafallopian transfer (GIFT)—offer hope to as many as 10–15% of the infertile couples who could not be successfully treated otherwise.[3] At least 100–120 medical teams in the US have established a record of some success with IVF, and proficiency with GIFT is increasing. However, some of the approximately 180 IVF/GIFT programs in this country have had no babies born to date.

As many as half of the infertile couples seeking treatment remain unsuccessful, despite trying various avenues of treatment. Knowing when to stop treatment is an individual matter for each couple. A decision often comes as couples ask themselves:

- Is further treatment worth the pain, expense, and disruption

- Is adoption or childlessness acceptable
- Is treatment costing so much that other goals are sacrificed
- If it is not yet time to stop, when will it be?

Couples seeking the most talked-about new reproductive technology, IVF, are often in a quandary over assessing practitioner skills. Is IVF experimental, or is it a proven medical therapy? At present, no blanket answer to that question is possible. Just as some physicians in IVF programs in the United States are proven practitioners of the art, others are as yet unproven; proficiency varies widely.

Costs of Treatment

How much does infertility cost? The dollar value of the personal, familial, and societal losses caused by infertility is inestimable. Americans spent, however, about $1.0 billion in 1987 on medical care to combat infertility.

Costs to individual couples receiving care for infer-tility vary dramatically, depending on the severity of their problem and their perseverance in seeking treatment. A complete diagnostic workup typically costs $2500–$3000, although many couples do not require such an extensive workup. Medical treatment may cost more than $22,000. Further, because conception is a precisely timed biological event, infertility diagnosis and treatment often involve the costs of time away from work, and may involve travel and lodging costs.

Surrogate Motherhood

Surrogate motherhood is more a social solution to infertility than it is a medical technology. About 15 surrogate-mother matching services are active in the United

States, and as many as 100 surrogate-mother arrangements may be concluded annually over the next several years. Legislation addressing surrogate motherhood has been introduced in more than half the state legislatures, and at least six states have passed laws either inhibiting or facilitating the practice. State court decisions have consistently found surrogate contracts to be unenforceable, even though they have differed on whether the contracts are legal.

Developing Technologies

Speculation about reproductive technologies yet on the horizon has captured the public's imagination like few other aspects of infertility treatment. The next decade will likely see the proliferation of the practice of embryo freezing as an adjunct to in vitro fertilization, or if success in freezing eggs comes about, there may be little need for most embryo freezing. Freezing eggs prior to fertilization, however, stands as a formidable technical task and may involve an insurmountable biological obstacle—damage to the fragile chromosomes of the egg.

Down the road a few years, the manipulation most likely to succeed is probably microinjection of a single sperm into an egg. Success here would mark dramatic progress in the treatment of male infertility, most of which is caused by too few or abnormal sperm. In IVF, 50,000–100,000 sperm are required to achieve fertilization; with microinjection, only one is needed. It is important to note that such an approach carries legal and ethical concerns: With microinjection, the clinician is selecting the one sperm that will fertilize the egg. This may present, for example, an interesting question of legal liability in the event of a congenital anomaly in the offspring.

Confronting Infertility

The reliable separation of X- and Y-bearing sperm to permit parents to select or change the odds of the sex of their baby remains elusive, despite many attempts. If and when sex selection of human sperm becomes possible, its widespread use will probably be limited by the willingness of couples to undergo the attendant artificial insemination or in vitro fertilization.

One technology of the present, in vitro fertilization, is itself a powerful means for unraveling mysteries of the human reproductive process. The advent of IVF permits researchers for the first time to view human reproduction in progress. Understanding the interactions between sperm and egg has potentially broad application not only for conception, but for contraception as well. Researchers in the area of human in vitro fertilization, however, have faced since 1980 the stifling effects of a *de facto* moratorium on Federal funding of research involving human IVF.

Policy Issues for Congress

In its report, OTA identified nine policy issues related to infertility that today confront Congress:

- Preventing infertility
- Reproductive research
- Reproductive health of veterans
- Information to inform and protect consumers
- Collecting data on reproductive health
- Providing access to infertility services
- Transfer of human eggs, sperm, and embryos
- Recordkeeping and
- Surrogate motherhood.

The first four of these policy issues have been a focus of interest by the US Government in the past year and are examined in the following discussion in greater detail.

Should Efforts Toward Prevention of Infertility Be Enhanced?

Preventing infertility is difficult. Factors that contribute to abnormal or too few sperm, for example, are largely unknown. Other factors, like endometriosis, are not amenable to prevention. Nevertheless, prevention strategies are desirable, because they may help some couples avoid the considerable emotional and economic costs associated with infertility treatment, and they may preempt some infertility that would be wholly untreatable.

Infertility resulting from sexually transmitted diseases (STDs)—an estimated 20% of the cases in the United States—is the most preventable. "With STDs," writes Dr. Willard Cates, Jr., of the Centers for Disease Control, "curative medicine equals preventive medicine."[4] The risk of infertility increases with the number of times a person has chlamydia or gonorrhea, the duration and severity of each infection, and any delay in instituting treatment.

There are an estimated 4 million chlamydia infections each year, many more than the number of gonorrhea infections (750,000 in 1987) and far more than the number of syphilis infections (30,000 in 1987). Yet, no national surveillance of chlamydia takes place, whereas gonorrhea and syphilis are required to be reported. A national system would allow the Centers for Disease Control and the various state health departments to identify high-risk groups and problem areas. Funds for screening and education could then be targeted to the appropriate populations and areas. The prevention and treatment of chlamydia that would result from these efforts would likely lead to lower rates of pelvic inflammatory disease (PID) and thus to decreased rates of PID-related infertility.

A national surveillance system would require state reporting laws or regulations. Although the Centers for Disease Control have consistently recommended that the states establish this surveillance system, individual states are unlikely to do so without additional funds. Congress is faced with the option of appropriating funds for the Secretary of Health and Human Services to make grants to state public health departments, thus helping them handle the costs of making chlamydia a reportable disease.

Beyond sexually transmitted diseases, there are many suspected factors contributing to infertility, but few confirmed culprits. Congress could direct the Secretary of Health and Human Services to establish a long-term research effort aimed at identifying exposures or behaviors in young adulthood that predispose an individual to infertility. Such long-term, longitudinal research that follows young adults through their reproductive lives is difficult, expensive, and often exceeds the research lifespans of individual investigators. In instances like this, therefore, coordinated, cooperative efforts (e.g., the Framingham Heart Study) are required. Such a study is critical for ferreting out confirmed from suspected factors contributing to infertility.

Without a comprehensive, longitudinal study to identify risk factors for infertility, many of them may never be fully defined, and possible preventive steps may never be taken. On the other hand, the result of such an undertaking may be confirmation that a number of cases of infertility are of unknown origin and not preventable.

***

Do Some Areas of Reproductive Research Need Additional Federal Support?

Research that involves fertilization of human sperm and eggs in vitro has since 1980 been excluded from federal support because of the absence of an Ethics Advisory Board within the Department of Health and Human Services. Such a board is required to advise the Secretary as to the ethical acceptability of such research (45 Code of Federal Regulations 46.204[d]). Without an Ethics Advisory Board in place, questions surrounding the interaction of sperm and egg—fundamental to an understanding of conception and contraception—remain largely uninvestigated.

On July 14, 1988, the Department of Health and Human Services announced its intention to reconstitute the Ethics Advisory Board.[5] With this announcement, the Department indicated its intention to abide by its own regulation requiring appointment of an Ethics Advisory Board.

Researchers indicate that they do not even attempt to obtain funding from the National Institutes of Health because of widespread awareness of the *de facto* ban on federal funding for IVF research. The dimensions of this chilling effect of the moratorium on IVF research are such that NIH estimated in 1987 that it might receive more than 100 grant applications related to human IVF if the Ethics Advisory Board were extant.

The effect of this moratorium on federal funding of IVF research has been to eliminate the most direct line of authority by which the federal government can influence the development of both embryo research and infertility treatment so as to avoid unacceptable practices or inappropriate uses. It has also dramatically affected the financial ability of American researchers to pursue improvements in IVF and the development of new contraceptives based on improved understanding of the process of fertilization.

Should the Veterans Administration Provide Infertility Diagnosis and Treatment?

The Veterans Administration (VA), the nation's largest health care delivery system, offers only limited treatment for infertility in its 172 medical centers and 227 outpatient clinics. On December 4, 1987, the Senate passed an amendment to Section 601(6) of Title 38 of the US Code that gave the VA authority to provide "services to achieve pregnancy in a veteran or a veteran's spouse where such services are necessary to overcome a service-connected disability impairing the veteran's procreative ability." The House did not pass such a provision, and in conference on April 13, 1988, the provision was dropped. A similar sequence of events took place in the 99th Congress.

The present position of the VA prevents it from treating infertility, since the agency does not interpret infertility to be a primary disability (defined as a disease, injury, or other physical or mental defect). Although some medical workup, such as varicocele repair in men or tubal surgery in women, may be performed, followup may be lacking. Under questioning by Rep. J. Roy Rowland at hearings of the Subcommittee on Hospitals and Health Care of the House Committee on Veterans Affairs on September 23, 1987, VA Chief Medical Director John Gronvall acknowledged that the sole functional way of assessing success of tubal surgery is to document successful fertilization in an unobstructed fallopian tube, and the VA is unable to provide that kind of followup to a female patient.

Procedures, such as artificial insemination, in vitro fertilization, and gamete intrafallopian transfer, are not provided by the VA. In addition, the VA lacks authority to treat a nonveteran spouse for infertility. This lack of authority is a problem indeed, because fertility is the product of interaction between two people, and so the infertile patient is, in effect, the infertile couple.

In 1985, about 16,000 male veterans and just over 1200 female veterans had known service-connected medical conditions that could lead to infertility. Among the men, the conditions range from removal of the testis or prostate to spinal injury. Among the women, the conditions ranged from removal of the ovaries to inflammation of the fallopian tubes or cervix.

Spinal-cord injury, caused principally by battlefield trauma during wartime, and vehicular and diving accidents during peacetime, is of special concern to both the VA (which supports 20 spinal-cord injury centers) and veterans' advocacy groups. The current outlook for fertility after spinal-cord injury in paraplegic men (although not women) is poor. Erection and ejaculatory dysfunction, compounded by infections of the urogenital tract, are common. Reproductive technology, in the form of electroejaculation and artificial insemination, offers a degree of hope to spinal-cord injury victims. Researchers at the University of Michigan Medical Center, for example, reported the first birth resulting from electroejaculation and in vitro fertilization.[6]

OTA was unable to develop meaningful cost estimates for provision of infertility services by the VA. Cost estimates will remain elusive until criteria are established for a number of variables. These include specification of the eligibility of veterans and/or spouses for infertility services, and the types of procedures to be provided. In addition, whether these services would be provided in-house or contracted to other facilities would greatly affect cost estimates.

Should the Federal Government Ensure that Consumers of Selected Infertility Services Have the Information to Make Informed Choices?

Congress generally does not regulate medical practice, with the exception of drawing broad criteria for care delivered at VA hospitals or reimbursed by federal insurance programs. Furthermore, medical techniques are not subject to consumer protection legislation, with the notable exception of Food and Drug Administration regulations for testing drugs and devices, and for regulating advertising of their indications and efficacy. Rather, quality assurance and consumer protection issues are left to state legislatures, professional societies, consumer groups, and word-of-mouth.

Short of regulating infertility treatment and research, Congress could exercise oversight to encourage NIH, for example, to hold a consensus conference on innovative infertility treatments. Such a consensus conference—of which more than 60 have been held in the last decade—could be used to:

- Influence the development of data collection on the use of IVF, gamete intrafallopian transfer, and other reproductive techniques
- Recommend indications for use
- Establish conventions for reporting successful outcomes and
- Define standards for laboratory equipment and personnel training.

In the area of consumer protection, Congress could direct the Federal Trade Commission to exercise its authority under Section 5(a) (6) of the Federal Trade Commission Act to examine whether advertisement of success rates at various IVF or gamete intrafallopian transfer clinics is misleading, and if so, to issue appropriate regulations. Regulations could be issued, for example, to standardize the ways in which success rates are reported, so that individuals are better able to make an informed choice about whether to undergo a procedure.

At a hearing held on June 1, 1988, the Subcommittee on Regulation and Business Opportunities of the House Committee on Small Business announced its intention to survey the services and success rates of IVF clinics in the United States.[7] To date, there has been no uniform public reporting of success rates at individual clinics in the United States.

Legislation Related to Infertility

The 100th Congress saw the introduction of more than a dozen bills related in some way to infertility (*see* Table 2). Whether permissive or inhibitory, the legislation taken as a whole recognizes that, for some couples, conceiving and forming a family has taken on added dimensions in the late 1980s.

Another index of lawmakers' interest in infertility is the breadth and number of congressional hearings—seven in the 100th Congress—that have dealt with the topic. Hearings in the US House of Representatives have been held by relevant subcommittees of the Committees on Government Operations, Small Business, Veterans Affairs, Post Office and Civil Service, Energy and Commerce, and the Select Committee on Children, Youth, and Families.

In addressing this issue, policymakers are subjecting to public discussion the kind of sexual topics generally consigned to closely held, private conversation. However, frank and open discussion is required if Congress is to treat infertile couples fairly.

Table 2
Bills Related to Infertility, 100th Congress

Bill	Sponsor	Description
House of Representatives		
H.R. 164	Guarini (D-NJ)	Provides federal grants to states to provide screening and diagnostic services for exposed individuals for health effects associated with diesthytilbestrol (DES)
H.R. 2220	Schroeder (D-CO)	Provides child adoption benefits for members of the uniformed services
H.R. 2221	Schroeder (D-CO)	Provides child adoption benefits for federal government employees
H.R. 2433	Luken (D-OH)	Prohibits certain arrangements commonly called surrogate motherhood
H.R. 2779	Waxman (D-CA)	Amends the Public Health Service Act to revise and extend the program of grants for the prevention and control of sexually transmitted diseases
H.R. 28	Schroeder (D-CO)	Provides that any insurer offering obstetrical benefits under the health benefits program for federal employees shall also provide benefits relating to certain "family-building" procedures
H.R. 3161	Montgomery (D-MS)	Provides services through

(continued)

Table 2 (continued)

Bill	Sponsor	Description
		the Veterans Administration to overcome service-connected disabilities affecting procreation
H.R. 3264	Dornan (R-CA)	Anti-Surrogate-Mother Act of 1987

Senate

Bill	Sponsor	Description
S. 9	Cranston (D-CA)	Provides services through the Veterans Administration to overcome service-connected disabilities affecting procreation **(Passed 12-4-87; dropped in conference, 4-13-88)**
S. 169	Reigle (D-MI)	Provides federal grants to states to provide screening and diagnostic services for exposed individuals for health effects associated with diesthystilbestrol (DES)
S. 268	Humphrey (R-NH)	Provides child adoption benefits for federal government employees, except for children conceived by artificial insemination, embryo transplantation, in vitro fertilization, or surrogate parenthood
S. 269	Humphrey (R-NH)	Provides child adoption benefits for members of the armed forces, except for

(continued)

Table 2 (continued)

Bill	Sponsor	Description
		children conceived by artificial insemination, embryo transplantation, in vitro fertilization, or surrogate parenthood
S. 2385	Kennedy (D-MA)	Amends the Public Health Service Act to revise and extend the program of grants for the prevention and control of sexually transmitted diseases, including chlamydia

References

[1] US Congress, Office of Technology Assesment (1988) *Infertility: Medical and Social Choices* (Washington, DC: US Government Printing Office).

[2] Hirsch, M. B. and Mosher, W. D. (1987) Characteristics of infertile women in the United States and their use of infertility services. *Fertil. Steril.* **47,** 618–625.

[3] Tagatz, G. E. (1989) Medical techniques for assisted reproduction, this volume.

[4] Cates, W., Jr. (1986) Priorities for sexually transmitted diseases in the late 1980's and beyond. *Sex. Trans. Dis.* **13,** 114–117.

[5] Windom, R. E., Assistant Secretary for Health, Department of Health and Human Services, testimony before the Subcommittee on Human Resources and Intergovernmental Relations, Committee on Government Operations, US House of Representatives (July 14, 1988).

[6] Ayers, J. W. T., Moinipanah, R., Bennett, C. J., Randolph, J. F., and Peterson, E. P. (1980) Successful combination therapy with electroejaculation and in vitro fertilization–embryo transfer in the treatment of paraplegic male with severe oligoasthenospermia. *Fertil. Steril.* **49,** 1089, 1090.

[7]Wyden, R., Chairman, Subcommittee on Regulation and Business Opportunities, Committee on Small Business, U.S. House of Representatives (June 1, 1988).

Sexuality and Assisted Reproduction

An Uneasy Embrace

Kathy J. Harowski

There is no doubt that mankind—only recently relabeled humankind—has long been engaged in a search for magic potions, and more recently technology, to change, improve, or eliminate sexual feelings, sexual behavior, and reproduction. History is full of examples of substances and mechanical contraptions used either to enhance or control sex and reproduction.[1] Everchanging, these inventions reflect the culture's sexual and reproductive goals and values, as well as the dominant cultural view of the role of men, women, and "experts" in the processes of sexuality and reproduction at a given point in time.

For example, the predominant themes regarding sexuality in the early nineteenth-century Victorian era revolved around the cultural "facts" that proper or decent women found sex morally repugnant and an acceptable activity only when procreation was the goal. Tannahill estimates that this myth and the various bans on intercourse during menstruation and pregnancy "imposed celibacy" on husbands and wives for approximately six of the first 12 years of marriage.[2] This lack of socially permissible sexual outlets for men was supported by the cultural theme of semen as a strength-giving, precious fluid that men should take steps to conserve for their physical and spiritual health.

Thus, a cultural norm of intercourse only for procreation, with an allowance of one time monthly seen as healthy and one time weekly for the "desperate" was advocated by Victorian experts on sex.[3] These same experts stated that sexual release—via intercourse or the vice of masturbation—that occurred more than once weekly would lead to certain debilitation and disease.[4] A technology to stop masturbation was the inevitable followup to these theories. The US Patent Office received many requests from individuals inventing antimasturbation devices.[5] As a culture evolves, it develops—and rejects—many norms about sexuality and reproduction. Occasionally, a sexual invention may find its niche elsewhere, as occurred with graham crackers and breakfast cereals, such as cornflakes, foods that were initially developed to decrease or eliminate sexual desire and behavior.[6]

Our culture presumes certain significant differences between current technological developments and the record of the past. First of all, the word technology is used rather than alchemy, magic, or Satan's work. We presume no more mystery or uncertainty, and report our results in terms of percentages and statistics. Our culture—and the medical model, a product of its times—views technology as a tool created to solve a specific problem or meet a specific need. So, in late twentieth-century America, we have sex therapists who effectively diagnose and treat all varieties of sexual dysfunction. Historically, any new technological developments produce positive and negative changes, raising the hopes as well as the fears and anxieties of those using the technologies.[7] This typically leads to an increasingly complex spiral of change, which creates more new technologies. From this point of view, the invention of the wheel might be credited with many of the joys and demons of our modern world. Medical technology in the past century has followed this pattern, with each advance creating new treatment opportunities, and frequently creating therapeutic dilemmas. This spiral effect has helped us arrive at our current level of sexual and reproductive technologies, which are quite powerful and offer real treatment alternatives for couples and individuals who consult the healthcare system with their sexual and/or baby-making difficulties.

The development of these powerful technologies also reflects a basic attitude shift in our culture regarding sexuality and reproduction. Sexuality is less an experience of romance, morality, or psychology, and much more an activity to be engaged in properly, with medical help to ensure that it is done correctly and, most recently, safely. Reproduction and pregnancy itself have undergone a similar transformation from "women's experience to medical subject."[8] The medical model, a creation of our current culture's values, has imposed notions of disease and science onto pregnancy, previously regarded as a natural process tended by women. Feminist authors emphasize the fact that medical and technological intervention in or control of pregnancy is a historically recent phenomenon that places women in a new role in the process of reproduction—a more mechanical, interchangeable one.[9]

A review of current media indicates that sexuality, reproduction, and parenting are cultural themes in transition. Tabloids, newspapers, and scientific journals discuss issues of safer sex, surrogate mothers, and test tube babies. A Twin Cities newspaper recently asked readers to send in "conception stories," which were printed in the feature section.[10] The majority of the stories centered around unusual locations for sexual behaviors that led to children with unusual names. The only story dealing with technology was that of one couple who conceived their last child eight days before the father's vasectomy. Although the general public is not uninformed about assisted reproduction, medical and surgical intervention in conception and pregnancy is not yet part of the common experience or expectation.

Typically, major cultural trends and shifts create confusion, uncertainty, and distress for individuals. Couples with reproductive difficulties in the 1990s may opt to use reproductive technologies while wistfully looking back to stories of making love on the beach and naming the baby Sandy. Although the technological advances of sexual medicine and assisted reproduction present important treatment options for individuals and couples, they often create complicated psychological dilemmas for those seeking help and for the healthcare professionals who work with them.

The Role of Technology in Sex Therapy

Sex therapists have a particular respect for technological innovation. Sex therapy has embraced new technology willingly; indeed, the field exists as an effective, focused method of intervention because of its alliance to technology. The pioneering work of Masters and Johnson[11] used available technology to assess and describe sexual functioning accurately in both men and women. When their book, *Human Sexual Response*, was published in 1966, it made available for the first time scientific data on a subject that had been considered mysterious and taboo. Masters and Johnson presented behaviorally focused treatments for sexual dysfunction, such as sensate focus and the "squeeze technique" for premature ejaculation, as the sex therapist's tools for resolving any sexual dysfunctions. These psychosexual "soft technologies" or structured processes[12] raised concern among more traditional psychodynamic therapists about the potentially damaging or distressing consequences of applying a powerful treatment technology to an individual or couple who might not be able to tolerate it for intrapsychic reasons. These concerns continue to be debated today among professionals in the field of sex therapy.

An even more complicated therapeutic situation occurs when we begin to consider hard technology: medical or surgical interventions for sexual dysfunctions. There are few studies of the effects of sexual technologies on psychological functioning and satisfaction. The research data in the sexuality field about the psychological ramifications of technological interventions for such conditions as erectile dysfunction, gender dysphoria, and genital reconstruction is limited and remains difficult to interpret.[13-15] An illustrative example is that of the penile prosthesis, a device surgically implanted in the penis to allow erection and sexual intercourse. These devices were developed, marketed, and used extensively in the last 15 years. Because they were used in response to clinical need and primarily by private practitioners, it remains difficult to assess postoperative complications and satisfaction with implants.[16] An additional concern, which highlights the differing interdisciplinary definitions of "successful" use of sexual technologies, is that the majority of studies conducted regarding implants inquired about surgical success, rather than sexual satisfaction or adjustment of the male and his partner.[17] In our current culture, male sexual performance pressure is a constant,[18,19] and some therapists question whether such medicalization of male sexuality—a perfect penis, surgically provided—provides a male the opportunity to experience his own feelings, needs, and expectations about sex genuinely.[20] We have even less information on the psychology of women and its interaction with sexual technologies.

Assisted Reproduction

The term "assisted reproduction" represents a rapidly growing and changing field of technologies, encompassing many different levels of medical involvement and intrusiveness in the process of conception. Interventions range from the kindly advice of a physician and home pregnancy tests to in vitro fertilization and surrogate motherhood. It is currently difficult to assess, particularly in any long-term sense, the psychological impacts of assisted reproduction on the processes of conception, pregnancy, and family formation. As was the case in the development of technologies for sexual dysfunction, clinical need has provided the impetus for treatment development, rather than models emphasizing research, ethical, or social perspectives. The marketing or selling of reproductive technologies raises additional ethical concerns. A recent Office of Technology Assessment (OTA)[21] report stated that approximately 50% of couples trying the various reproductive techniques conceived successfully and that "quality of care varied widely." This report states further that sexual behavior may be "radically" changed in couples seeking pregnancy and that counseling is an "often underutilized component of infertility treatment." Thus, a medical model has been applied to an intense interpersonal experience, with limited technical success and a lack of support for the emotional/psychological reactions to the process and outcome. Understanding the impact of reproductive technologies on the psychology and sexuality of the couples and individuals who seek to use it is a complicated process in the current climate.

The Desire to Parent

The wish to become a parent is multidetermined for both males and females, with biological, social, and psychological factors being important contributors. Gender role and the history of the client's family of origin are additional components in the decision for the individual to parent. Each individual brings their unique perception to conception and parenting—traditionally, two together create a life and a family. The wish to parent remains complex, whereas the other steps in the process—conception, pregnancy, and child rearing—can be modified with technology and are no longer available only to fertile, sexually active, heterosexual couples.

Attempting to understand the meaning of parenthood, reproduction, and sexuality for the individuals and couples with whom we work is a great challenge to the healthcare worker. Our current conceptualizations of parenting attempt to define both biological and psychological parameters by talking about motivation, cognition, and attachment in attempts to understand why—or why not—couples or individuals act on the biological possibilities of sexuality and reproduction. Sociobiologist Melvin Konner describes a visit to the pediatrician with his colicky six-week-old daughter in the following way:

> ...I held the baby up to the light, squinted at the physician out of one bloodshot eye, and made my

> statement..."She's ruining my life. She's ruining my sleep, she's ruining my health, she's ruining my work, she's ruining my relationship with my wife, and... and...and she's ugly...Why do I like her?"
> The physician, a distinguished one in our town, and a wise and old and virtuous man seemed unbaffled by the problem. "You know"—he shrugged his shoulders—"parenting is an instinct and the baby is the releaser."[22]

Rather than describe the human urge to parent as a simple instinct, sociobiologists view it as a tendency or possibility built into humans, "drawing upon a deep well of ancient, stereotyped emotion, thought, and action" stored in the nervous system and the genes.[23] This well is tapped for many reasons, ranging from ensuring reproductive success and survival of the parent's genetic material to the psychological processes of attachment and affection.

It was John Bowlby who first decribed the process of attachment between a child and its primary caretaker that forms the basis for all later social, sexual, and affectional behavior. Bowlby's three-volume work, *Attachment*,[24] *Separation*,[25] and *Loss*,[26] details the complexity and ongoing nature of this process, which creates the need and capacity for love in all its adult forms, including parent–child.

Powerful messages about gender and sexuality are also related to the wish to be a parent. Individuals who have difficulty reproducing will often label themselves as defective or impaired men or women. Often, the gender-related, sexual performance pressures males and females experience become clear when conception is elusive. For example, men feel extreme pressure to "perform sexually"—to have an erection, use it in a way that will please their partner, maintain it for an adequate amount of time, and so forth. Sex therapy models emphasize relaxation and focusing on one's own pleasure as antidotes to the fantasy of having a penis that is "two feet long, hard as steel, and can go all night."[27] Being informed that his sperm count is low creates even more anxiety for a male who has been acculturated to believe he must be sexually perfect and virile to be a successful male. For most women, motherhood is a cultural and social role that is still an expectation rather than an option. Any interference in obtaining the desired outcome of being a mother will likely affect self-esteem, body image, and sense of femininity. Women describe a sense of loss, emptiness, anger, and incompetence with infertility.[28]

Finally, family of origin plays an important role in the desire to parent, in terms of history and role modeling. As a therapist, I have learned that a very powerful question to ask couples is, "What does it mean to be a mother/father?" This simple inquiry and the replies it generates can form the basis for an entire family history, and elicit important intergenerational myths and patterns in parenting. After stating and examining their historical baggage, couples and individuals can create meanings of their own for parenting—and try to undo the past if unhealthy patterns are present.

The meaning of reproduction within a relationship represents the marriage of the views of the individuals involved. Models of relationship/marital therapy have become increasingly sophisticated in assessing how individual psychodynamics operate in family systems. When a couple determines that they are "ready" to have a child, sexual patterns change to achieve the desired result: pregnancy. If successful, the purposeful sexual activity becomes a pleasant memory; if conception does not occur, sexual pressure replaces pleasure.[29] In examining conception from an object relations perspective, Scharf and Scharf state: "Conception becomes a higher priority than spontaneous sexual intercourse. Not yet conceived, the child is dominating the couple as they feel longing to have the child yet feel spurned by it."[30] Again, each partner will have different levels of motivation and reaction to these stressful circumstances. How a couple copes with the conscious and unconscious reactions to a difficult conception may determine their future together. The involvement of others—or technology—in the process adds to the potential for conflict.

Humans and Biomedical Technology

Numerous studies document the profound psychological effects that occur when humans and technology are brought together to prolong or enhance life. Healthcare workers have documented this phenomenon in patients who struggle with the psychological implications of kidney dialysis, rehabilitation from spinal-cord injury, and organ transplants.[31,32] On a less complex note, the issues of denial and ambivalence found in many diabetics who work to accept their need for insulin, and the compliance issues in conditions as diverse as hypertension and psychotic disorders, are also well documented.

Extrapolating these findings to a discussion of sexual and reproductive technology would seem to place one on safe, if uncharted, ground. Sex therapists have long advocated the involvement of both partners in any decision-making regarding technology, to reduce anxiety and decrease the potential for iatrogenic (treatment created) conflict. For example, a man insisting on a penile prosthesis to surprise his partner with a return to erectile and intercourse capability may undergo an invasive, expensive surgical procedure for no reason if his partner is secretly relieved that they no longer have sex. Women undergoing surgical treatment of genital or vaginal cancer report that, without truly informed consent, vaginal reconstruction has led to disappointing sexual outcomes for themselves and their partners.[33]

Assisted Reproduction and Sexuality

Our culture's sexual experts have abandoned describing or prescribing sexual behavior in terms of norms or averages. Therapists now focus on helping couples and individuals obtain the new ideal—relaxed, orgasmic sex between two mutally consenting equals. The cultural expectation is one of spontaneous sex without anxiety or performance pressure, labeled by Masters and Johnson[34] as the most potent interruptor of sexual functioning.

Consider partners seeking pregnancy. What was a spontaneous, perhaps capricious event is now timed for "success." Pressure builds if the process of conception does not occur as planned. Men and women alike describe a sense of objectification occurring around sexuality, as the pressure to conceive increases. Couples in therapy report less pleasure and an increased sense of anxiety about sex under these conditions. One male reports rushing home from work during his lunch hour to have intercourse with his wife in spite of a head cold and a traffic jam. A high-powered professional woman whose job required frequent travel recalls tearfulness and a sense of grieving for lost opportunity when noting body changes associated with ovulation while out of town. Clinical populations report an increase in sexual dysfunction in response to these pressures.[35]

The existence of assisted reproduction technologies may create specific areas of conflict for individuals, couples, and the healthcare professionals who work with them. One potential area of conflict is within the individual, who may begin the process of assisted reproduction quite willingly and without anxiety, only to find that unanticipated feelings and conflicts appear. For example, a male agreeing to artificial insemination by donor may find himself becoming angry and withdrawn because he feels like a nonparticipant in his partner's pregnancy. Several surrogate mothers, including Mary Beth Whitehead Gould, have experienced conflict and uncertainty about their role, whether during the pregnancy, immediately afterwards, or years later.[36]

Another arena for psychological conflict exists within the relationship or couple system. A partner who is less committed to the notion of parenting may balk at the intrusiveness, expense, and psychological stress of assisted reproduction, setting the stage for marital dysfunction. In terms of extended families, intergenerational boundaries and issues of confidentiality often become a factor. Couples may disagree on whether or not extended family members should be told about or participate in reproductive interventions.

Interdisciplinary misunderstandings and conflicts may arise from the collaboration across healthcare professions because of the differing treatment goals, perspectives, and professional incentives of the different professions. Team members focusing on mental health may find themselves in sharp disagreement with those working from a more traditional medical model. One couple decided to abandon psychotherapy in favor of insemination to achieve pregnancy because of the sexual aversion experienced by the male partner. This created much debate among the treatment team regarding the use of technology to avoid or bypass a significant intimacy and sexual dysfunction. As previously noted, counseling is frequently underutilized in couples undergoing treatment for infertility.[37]

The search for a perfect baby is a current cultural preoccupation. Biotechnology for reproductive purposes creates a sense that soon couples will be able to create a child to order: gender, genes, eye color, and so on, but what about couples who achieve pregnancy only to be told that the fetus shows nonspecific genetic abnormalities on amniocentesis? What does it mean to the couple—and to the diversity of the human race—to abort in this situation? Gender preference for first-born males means that a majority of parents would choose a male child as their first born. Masters, Johnson, and Kolodny[38] highlight the implications to women—and to society—of the majority of females being "second born." Would this mean lower occupational status and self-esteem for women, reversing recent trends towards equality?

Suggestions for Future Discussion and Research

Newman[39] highlights the need to help couples or individuals think about the reproductive process in which they are involved and recommends help for all involved—both professionals and consumers—in understanding the process. Potential parents need to create a framework for assisted reproduction in terms of their own myths and history; a technological parable comparable to the story of having sex on the beach and naming the baby Sandy. All of those enmeshed in the process need a framework to help them deal with the intense psychological reactions to creating life. In particular, Newman[40] emphasizes the need to help couples decide when and how to stop if they are not meeting with success, a concern echoed by the OTA report.[41]

The availability of reproductive technologies raises questions about the ethics of their use, especially around the issue of iatrogenic conflict—creating turmoil or conflict via the use of technology. Clinicians must emphasize the importance of individual evaluation and assessment to determine whether or not a technological solution will be a positive option. In particular, we must ask the "What if" questions of ourselves and the individuals with whom we work, aiding all involved in the decision-making process to assess the immediate benefits and the long-term positive and negative effects of any technological intervention.[42] Our culture is just beginning this debate in the courts and the media; healthcare providers have an obligation to contribute our data. The power of reproductive technologies is especially clear when talking to the people who can be described as the mistakes of such assisted reproduction. In this sense, I do not mean the children produced, but the parents—surrogate, birth, or social—or families who either are not satisfied or for whom the procedure creates such great conflict that their lives are negatively impacted.

Masters, Johnson, and Kolodny, in their book *Sex and Human Loving*, address the "sexual impact of biotechnology" by describing a brave new world in which surrogate pregnancies are sought and artificial wombs created to spare women the hazards of child bearing. Although they state that assisted parenting options will reduce the marriage rate, they do not discuss the ethical issues involved in using such technologies. They end this book with the hope that better birth control methods will increase the emphasis on sex for pleasure.[43]

Healthcare practitioners have a responsibility to help patients understand their reactions to and feelings about the process of reproductive technology. Maintaining this as a goal may help all the healthcare professions avoid a medical-model approach to sexuality and reproduction. This approach pushes for either/or solutions, might insist on or create financial incentives for either/or solutions, and views reproductive concerns as either organic or psychological. An either/or attitude would ignore a substantial body of literature, as well as experience, indicating that where sex and reproduction are concerned, there is always much overlap between medical and psychological factors. The need for continued psychotherapeutic support and therapy when technological intervention is employed is becoming clearer as our body of experience grows.

In the long run, the availability of technology and the options it presents will be of great benefit to the field of sexology and the study of reproduction. Technology will help us to understand more fully the psychology of sex and the desire to parent, because its use will help us to further operationalize and define our terms. The development and proliferation of sexual technologies emphasize the wisdom of the pioneers in the field of sexology who sought to create the first truly interdisciplinary study of human functioning. Assisted reproduction, with all of its blessings and dangers, makes even more clear the need for study and collaboration among many professions. This is most important when contemplating the design and execution of long-term followup studies of families created via assisted reproduction.

References

[1] Tannahill, R. (1981) *Sex in History* (Stein and Day, New York).
[2] Tannahill, R.
[3] Tannahill, R.

[4]Money, J. (1985) *The Destroying Angel: Sex, Fitness, and Food in the Legacy of Degeneracy Theory, Graham Crackers, Kellogg's Corn Flakes, and American Health History* (Prometheus, Buffalo, NY).
[5]Tannahill, R.
[6]Money, J.
[7]L'Allier, J. (1987) *Right Answer Wrong Question. Focus on Teaching and Learning* (University of Minnesota Press, Minneapolis, MN).
[8]Oakley, A. (1987) From walking wombs to test-tube babies, in *Reproductive Technologies: Gender, Motherhood, and Medicine* (Stanworth, M., ed.) (University of Minnesota Press, Minneapolis, MN).
[9]Stanworth, M., ed. (1987) *Reproductive Technologies: Gender, Motherhood, and Medicine* (University of Minnesota Press, Minneapolis MN).
[10]*Star Tribune: Newspaper of the Twin Cities*, "Great Beginnings," compiled by Peg Meier, staff writer (May 6, 1988).
[11]Masters, W. and Johnson, V. E. (1966) *Human Sexual Response* (Little, Brown, & Co., Boston).
[12]L'Allier, J.
[13]Cairns, K. and Valentica, M. (1986) Vaginal reconstruction in gynecologic cancer: A feminist perspective. *Journal of Sex Research* **22(3)**, 333–346.
[14]Lothstein, L. (1982) Sex reassignment surgery: Historical, bioethical, and theoretical issues. *Am. J. of Psychiatry* **139**, 417–426.
[15]Tiefer, L. (1986) In pursuit of the perfect penis: The medicalization of male sexuality. *American Behavioral Scientist* **29(5)**, 579–599.
[16]Tiefer, L.
[17]Collings, G. and Kinder, B. (1984) Adjustment following surgical implantation of a penile prosthesis: A critical overview. *J. Sex Marital Ther.* **110**, 225–271.
[18]Masters, W. and Johnson, V. E.
[19]Zilbergeld, B. (1978) *Male Sexuality: A Guide to Sexual Fulfillment* (Bantam Books, New York).
[20]Tiefer, L.
[21]US Congress, Office of Technology Assessment (1988) *Infertility: Medical and Social Choices* (GPO, Washington, D.C.).
[22]Konner, M. (1982) *The Tangled Wing: Biological Constraints on the Human Spirit* (Holt, Rinehart, & Winston, New York).
[23]Konner, M.
[24]Bowlby, J. (1969) *Attachment* (Basic Books, New York).

[25]Bowlby, J. (1973) *Separation: Anxiety and Anger* (Basic Books, New York).
[26]Bowlby, J. (1980) *Loss: Sadness and Depression* (Basic Books, New York).
[27]Zilbergeld, B.
[28]Office of Technology Assessment.
[29]Scharf, D. and Scharf, J. (1987) *Object Relations Family Therapy* (Routledge and Kegan Paul, London).
[30]Scharf, D. and Scharf, J, p. 90.
[31]Finkelstein, F. O. and Steel, T. E. (1978) Sexual dysfunction and chronic renal failure: A psychosocial study of 77 patients. *Dialysis Transplantation* **7(9)**, 877–923.
[32]Woods, N. F. (1984) *Human Sexuality in Health and Illness*, 3rd Ed. (C. V. Mosby, St. Louis, MO).
[33]Cairns, K. and Valentica, M.
[34]Masters, W. and Johnson, V.E.
[35]Menning, B. (1977) *Infertility: A Guide for the Childless Couple* (Prentice Hall, Englewood Cliffs, New Jersey).
[36]Office of Technology Assessment.
[37]Office of Technology Assessment.
[38]Masters, W., Johnson, V. E. and Kolodny, R. (1986) *Sex and Human Loving* (Little, Brown, & Co., Boston).
[39]Newman, L. (March, 1988) New Reproductive Technologies: Framing the Ethical Issues, (unpublished) paper presented at the Society for Sex Therapy and Research Conference, New York, New York.
[40]Newman, L.
[41]Office of Technology Assessment.
[42]L'Allier, J.
[43]Masters, W., Johnson, V.E. and Kolodny, R.

Arguing with Success

Is In Vitro Fertilization Research or Therapy?

Arthur L. Caplan

The First Phase of the Debate in the Evolution of IVF—Do No Harm!

In the 1950s and 60s, in vitro fertilization (IVF) was viewed by even its staunchest advocates as experimental.[1] No children had been produced using the technique. Serious doubts existed about whether IVF could be used to create a baby and whether any baby that resulted would be healthy. Those involved in extending IVF techniques from animals to humans worried well into the 1970s that "the normality of embryonic development and the efficiency of embryo transfer cannot yet be assessed."[2]

Critics of IVF also viewed the technique as experimental. Their primary concerns were that it was immoral to risk harm to children by using the technique; that it was disrespectful of the humanity of gametes and embryos to experiment upon them, since researchers viewed them as nothing more than "mere stuff;" and that it was immoral to conduct research on subjects incapable of giving consent.[3]

This last objection was the most potent. It actually came in two forms. One line of argument held that research on embryos was immoral, since the people who might be created could not consent to accept the risks involved. Another version of the objection held that the embryos involved in research to develop IVF might be at risk of malformations or other physical harms, but could not consent to accepting such risks. In both versions of the argument, the immorality of research on unconsenting subjects played a key role.

Paul Ramsey and Leon Kass argued that, if there were risks to subjects who lacked the ability to consent, then IVF "constituted unethical medical experimentation on possible future human beings" and, as such, ought to be "...subject to absolute moral prohibition."[4] They argued that the possibility of doing irreparable damage to an embryo or to a potential future person, neither of whom could consent to serve as an experimental subject, was a sufficient ethical basis for prohibiting research on IVF techniques.[5]

Persuasive arguments have been made against the view that any and all forms of risky research upon nonconsenting subjects are morally illicit.[6] Yet, these rejoinders were not all that influential in the debate about IVF.

Debates about the morality of experimentation involving IVF ended when a team of British doctors succeeded in creating a baby who appeared to be quite normal using IVF techniques. There were a few efforts[7] to keep open the issue of experimentation using unconsenting subjects, but the debate over the ethics of the procedure quickly shifted to other topics.

Phase Two of the Debate— You Cannot Argue with Success

The birth of Louise Brown in 1978 had an enormous impact on the content of the subsequent ethical debate concerning the use of IVF. Earlier ethical arguments presumed that the use of IVF to treat infertile women was clearly experimental. Whereas one might condemn the decision to use IVF to produce a baby, the appearance of Louise Brown, along with a few thousand more like her in the United Kingdom, Australia, the United States, and other Western European countries, effectively ended all arguments about both efficacy and safety. As a result, those pursuing IVF were able to redescribe what they were doing. Once Louise Brown had been born, IVF was no longer experimental; it was therapeutic.

Post-Louise Brown, ethical debates about reproduction by means of IVF underwent a drastic change. Critics began to question the morality of procreation and conception occurring outside the context of heterosexual intercourse.

For example, the recent Vatican instruction[8] on assisted reproduction expresses no concern about the experimental nature of IVF. Instead, the unnaturalness of separating sexual intercourse and procreation, and the threat IVF poses to the stability of the nuclear family are the bases for the condemnation of IVF as morally illicit.

Similar criticisms are evident in the more recent writings of Leon Kass. He notes in the course of a discussion of the ethics of ectogenesis that it is "incompatible with the kind of respect owed to its [the embryo's] humanity that is grounded in the bonds of lineage and the nature of parenthood."[9]

The problem with ethical arguments that locate the moral illicitness of IVF in the unnaturalness of the procedure is that they are difficult to take seriously unless one is also willing to reject key elements of modern obstetrics. No evidence exists that the use of glass rather than protein as the medium to facilitate conception has any adverse impact on the children who have been conceived in this way, their parents, or the viability of the family as a social institution. Americans do not think less of those whose entry into the world was assisted by such unnatural means as forceps or a Caesarean section.

Natural origins are of ethical interest only to the extent to which societies attach stigma or opprobrium to other methods of creation. A society may decide to view medically assisted conception as morally irrelevant to the dignity and worth of human beings in the same way that no opprobrium attaches to those born in hospitals rather than homes. There is no self-evident reason to presume that there is anything morally illicit about a human being originating via extracorporeal fertilization.

Forces Driving the Evolution of Experiments into Therapies

It is extremely interesting to note that the appearance of Louise Brown and many hundreds of other healthy infants was seized upon by proponents of IVF to remove the labels of experimentation and research. This should not be a surprise. It was difficult for those who wanted to conduct the research that might result in a "Louise Brown" to counter the charge that they were experimenting on incompetent and helpless subjects. Whether or not a case could be made for putting embryos and future persons at risk, the concern about the morality of using in-

In Vitro Fertilization

competent subjects in risky research had moral force. The best tactic was to remove the label of experimentation from IVF.

The desire of biomedical scientists who do human experimentation to shed the label of experimentation quickly has interesting analogs in other areas of moral dispute concerning medical innovation. Those involved in attempting the first artificial heart implant and who conducted the first xenograft of a heart to a young infant were quick to declare their efforts therapeutic and non-experimental. The therapeutic status of these interventions was seen as established by the demonstration of the mere feasibility of the undertaking.[10]

There are many reasons why those involved in the development of new medical interventions wish to dispose of the label "experimental" as soon as they can. Experimentation carries with it connotations of the unknown, the risky, and the especially dangerous. Talk of experimentation can make it difficult to recruit willing subjects. Such connotations are unfortunate, since there are many experiments that are relatively safe and risk free, and many therapies that are precisely the opposite.

Another reason for the rapid transmutation of experiments into therapies is hope. Those who treat the sick and those who are sick or disabled naturally want to have a cure. Talk of "therapy" is far more conducive to optimism than is talk of "experimentation" or "research."[11] Researchers involved in the development of new drugs to treat terminal cancers or to help those suffering from AIDS find it much easier to offer comfort to the dying using the language of therapy, rather than that of research.

Another factor influencing the speed with which experimentation becomes therapy is the assignment of credit for discoveries. Few scientists or physicians get credit or public acclaim for being the first to conduct experiments. Credit goes to those who find cures, who discover therapies. If the first artificial heart implant is still experimental, or the first baby born as a result of IVF constitutes research, then others may try to claim credit for the first truly "successful" application of a new technology or technique. The race in biomedicine for fame, fortune, and celebrity goes to those who find cures. Priority is no small matter in a scientific world of fierce competition for grants, fame, and recognition.

There is another important force driving new techniques, drugs, and devices down the continuum from research to therapy—money. Third-party payers, whether public or private, do not want to pay for experimentation. Those giving grants for basic research do not want to fund therapy.

Clinical researchers, those who make the first efforts to try new techniques, drugs, or devices on human beings, are caught in a very uncomfortable bind. If they say what they are doing is research, then those they treat may not be able to obtain reimbursement for their costs from their insurers. If they describe what they are doing as therapy, then agencies involved in funding research will discontinue funds.

Faced with this dilemma, many researchers choose to dispose of the label of experimentation.[12] This choice permits their subjects to gain relief from what are often overwhelming costs. It also permits access to a larger pot of funds than would be available by sticking with the experimentation description, since at least in the American context, funding for therapy is far more generous than funding for research.

Currently, in the United States, most third-party payers, public and private, do not cover the costs of IVF.

Five states have enacted legislation requiring the costs of IVF to be included in insurance plans.[13] The unwillingness of third-party payers to reimburse the costs of IVF throughout its evolution has led to frequent and solemn invocations of the language of therapy in descriptions of IVF at federal and state legislative hearings and in the media.

Those involved with the development of IVF had strong reasons to want to rid the technique of any association with experimentation. Their critics were persuasive in arguing that experimentation ought to be permitted only on those capable of understanding and consenting to it. Young children, and certainly embryos, cannot give informed consent. It was hard to justify ethically continued efforts to develop IVF in the face of these moral concerns.

Once Louise Brown appeared, was photographed, hoisted about, and otherwise exhibited as a paragon of pediatric normality, the thrust of the ethical worry about IVF could be deflected by putting to rest the sticky matter of the acceptability of undertaking risky research on the unconsenting. Therapy upon the unconsenting is much less morally vexing.

What Criteria Ought Govern the Language of Experimentation and Therapy?

It is interesting to see exactly how existing regulations governing human experimentation define research. The so-called Belmont Report, published in 1979, which played a key role in the formation of existing federal guidelines concerning human experimentation, used the following definitions:

practice—interventions designed solely to enhance the well-being of an individual patient that have a reasonable expectation of success

research—an activity designed to test a hypothesis, permit conclusions to be drawn and thereby to develop or contribute to generalizable knowledge[14]

Existing regulations governing institutional review boards (IRBs) reflect the considerations raised in the Belmont report definitions. Research is defined as, "a systematic investigation designed to develop or contribute to generalizable knowledge."[15]

These definitions place great weight on intent. If a biomedical scientist believes that what he or she is doing is done solely to benefit a patient with some chance of success, then what is done is a part of practice. It is therapy. If the goal or intent is to produce generalizable knowledge, then what is done is research.

However, these definitions leave too much to intentions. Although the goals of therapy and research clearly are different, it is odd to make the distinction entirely contingent upon the aim of the healthcare professional. Even the most honest and forthright clinician is going to have a hard time describing what he or she is doing as research if it means adverse and often disastrous financial consequences for patients.

Whatever the defining characteristics of research, they must go beyond the subjective intent of the healthcare provider or scientist. Two characteristics that would appear to be especially relevant are the state of knowledge prevailing about the underlying mechanisms or processes that produce a particular result, and the efficacy associated with a particular activity in terms of the probability that it will produce an intended outcome.

If a physician can produce a cure in a patient, but does not have any idea why the cure comes about, then, in a key respect, the intervention is experimental. The reason for this is that it may not be clear exactly what intervention or contributing factor is responsible for producing the outcome that is sought.

For example, dermatologists have long known that a combination of ultraviolet light and coal tar helps many persons suffering from psoriasis. The cure rate associated with this regimen is quite high. Nonetheless, physicians remain uncertain of exactly which frequencies of ultraviolet light, which components of the coal tar, and which combinations of light and coal tar are responsible for symptomatic relief. The treatment is certainly useful. However, it is still experimental. Why it works is at least as relevant to the description of what dermatologists do as the fact that it has benefits for those who receive the ministrations.

Similarly, interventions that simply fail to bring about intended results are experimental even if those who use them intend only to help their patients. Oncologists may describe various drugs as therapeutic for lung cancer, cancer of the pancreas, or cancer of the liver, but the fact is that no available drug regimens are efficacious against these forms of cancer. Whereas an oncologist may believe he or she is providing therapy to those with cancers of this sort, the fact is that what is done is much more appropriately described as, at best, experimentation.

Background knowledge concerning both causal mechanisms and efficacy is a key factor that must be considered in assigning a particular intervention a place on the experimentation–therapy continuum. Intentions in and of themselves are not sufficient to distinguish research from therapy.

Is IVF Really a Therapy?

More than 3000 babies have been born by means of IVF since the appearance of Louise Brown in 1978. Optimism and declarations of success attend many public and even private scientific discussions of IVF. Moreover, the proliferation of American clinics and hospitals offering IVF since 1980, when the first American clinic opened its doors, has been rapid enough to persuade even a casual observer that IVF is a technique that works, regardless of the moral reservations some may feel about it. Yet these facts do not convey the whole story with respect to IVF. Even if one looks to the most charitable presentations of the success rates associated with standard IVF, the figures are not impressive.

For example, the Ethics Committee of the American Fertility Society (AFS) states that:

> Success rates with human IVF techniques have steadily improved since the first birth in 1978. Currently, the success rates vary but have been reported to be up to 25% per cycle of treatment.[16]

The 25% figure in and of itself is especially impressive, since it can be interpreted as showing that all persons who use IVF techniques can expect to have a baby at the end of four IVF cycles.

> The Medical Research Council/Royal College of Obstetrics (MRC/RCOG) and Gynecology are even more glowing in their assessment of the clinical status of IVF today:
>
> ...it should be regarded as a therapeutic procedure covered by the normal ethics of the doctor/patient relationship.[17]

However, the AFS and MRC/RCOG claims are deceiving. It is true that some clinics have reported rates of success per cycle in the range of 25%, but these are atypical. Many American clinics report no statistics, succesful or otherwise. They are under no compulsion or requirement to do so. Of those that do, few have success rates anywhere close to 25% per cycle. One recent study estimates that fewer than half of the 169 centers now offering IVF have ever created a baby.[18]

Those seeking information on the success rates associated with IVF are almost always given data based upon the rosiest self-reports of those facilities that choose to make their experience with IVF public. No data exist on the ability of individual doctors to successfully use IVF. Nor are rates for individual centers available, since there are no standardized registries to which efficacy data must be reported. Nearly all of the clinics that do report data do so only as part of a pool of clinics, making it impossible to associate particular rates with either individual doctors or specific clinics.

Another problem with such statements as that of the AFS is that success can be defined in many ways. Success may be defined in terms of embryos created, implantations, clinical pregnancies, pregnancies confirmed by determining a fetal heartbeat, total number of babies born, or number of couples who have gone home with a baby.[19]

To those seeking IVF, it is the take-home baby rate that counts. No one seeks IVF simply to produce an embryo or a pregnancy. Babies are and must be the final standard by which the technique is assessed.

If it is true that it is the number of babies who go home from the hospital that is the figure that counts in assessing the efficacy of IVF, then there are numbers that cast the rosy statements of the AFS and the MRC/RCOG into some doubt. A recently published report[20] based on the experience of the 30 centers offering IVF in the United Kingdom in 1985 reveals the following outcome information:

Total patients	2706
Total tx cycles	3257
Total conceptions	379
Babies	289

These numbers indicate that those seeking IVF in the United Kingdom in 1985 had a less than one in ten chance of obtaining a baby from a single IVF cycle. The less than one in ten chance per cycle does not say anything about how many children were born prematurely or with low birth weight. These are complications that are associated with those forms of IVF that utilize multiple embryo implants and thus run the risk of producing multiple pregnancies.

American statistics are not any more reassuring on the matter of IVF and its connection with babies. Data voluntarily reported[21] from 41 American clinics for 1986 show the following:

Total patients	3055
Total tx cycles	4867
Total conceptions	485
Babies	311

About 10% of American couples who receive care at one of these 41 clinics can expect to become pregnant. Of these, about 65% have a baby. So the rate of babies per treatment cycle is just under 7%.

There are no clear-cut rules for calling a technique experimental or therapeutic. From the point of view of an infertile couple who succeed in having a baby using IVF, their success is sufficient to establish IVF as a therapy.

From the point of view of the individual IVF provider, the intent to create a baby suffices to establish IVF as therapeutic.

Yet from a broader perspective, it is hard to imagine using the label of therapy for a procedure with a success rate of less than one in ten per trial. The efficacy of IVF, despite the fact that it has been used to produce thousands of babies, is still very, very poor.

IVF success rates per cycle are poor. However, it is fair to ask, poor compared to what? After all, engaging in reproduction in the old-fashioned way results in babies only about 10–15% of the time. Still, such an argument misses crucial differences between having babies as a result of sexual intercourse and having them whipped up in a dish.

The costs of IVF are not small. Per cycle charges in American fertility clinics range from $4000–10,000.[22] Also, the process of egg retrieval, drug treatments, and intensive monitoring by various specialists is not a matter of indifference to the couples who receive IVF.[23,24] Most importantly, declarations that IVF is a therapy obscure the fact that still relatively little is known about underlying mechanisms of fertilization, implantation, and development.

How Well Is IVF Understood?

Not only are the rates of success of standard IVF low, but in addition ignorance about why this is so is quite high. Key components of IVF are not well understood. The proper components of the media in which sperm and egg are joined, the dosage levels for various hormones to produce superovulation, the best temperatures and environmental circumstances for storing and reimplanting embryos, which embryos are most likely to implant, and other basic questions are still shrouded in a fair degree of mystery.[25]

IVF is, at present, still more empirical than scientific. There is more trial and error than systematic inquiry present in the manner in which IVF is carried out. If background knowledge is relevant to the decision about whether an intervention is still experimental, then IVF would seem to still be on that side of the ledger.

Moreover, the chance of undertaking the kind of experimental work that would provide answers to some of the underlying mechanisms of reproduction is quite low. The failure to view IVF as experimental, or at least highly innovative, leads to a regrettable indifference on the part of many in the field to the formulation of adequate and comprehensive international data bases, the tolerance of less than high-quality performance by some practitioners and clinics, and the inability to engage the public in a rational inquiry into the question of whether the time has yet arrived for public funds to be allocated for the support of research into the processes involved in reproduction.

There has been no federal money allocated to basic research in the area of assisted reproduction since 1978. Various politicians have found the prospect of wrestling with the morality of experimentation on embryos, eggs, and sperm too difficult to address. This failure of leadership has created a situation in which more than 200 IVF centers exist, but almost no American scientists are trying to understand how IVF works.

In the short run, the talk of therapy may have served to deflect moral criticism of early IVF programs. In the long run, it deflected public policy away from the hard choices that needed and still need to be addressed in order to refine and perfect the technique.

Reproductive Rights and Liberties

Information about the efficacy of IVF is crucial to both consumers and payers if they are to make reasonable decisions about whether or not to utilize the technique. However, it might be argued that, whatever the experimental or therapeutic status of IVF, the government has no proper business attempting to restrict access to or regulate the technique.[26]

American courts have been concerned to protect procreation from outside interference by third parties. The Supreme Court held in *Eisenstadt v. Baird* that

> if the right of privacy means anything, it is the right of the individual, married or single, to be free of unwarranted governmental intrusion into matters so fundamentally affecting a person as the decision whether to bear or beget a child (405 US 438, 453 [1972]).

The report of the Ethics Committee of the American Fertility Society, *Ethical Considerations of the New Reproductive Technologies,* cites this case in arguing that the right of privacy vis à vis reproduction implies that IVF and related methods for assisting reproduction "...may not risk sufficient tangible harm to the parties or the offspring to warrant state interference with the constitutional right to procreate."[27] The report concludes that:

> [i]ntangible religious, moral or societal concerns about the nature of reproduction, family, the reproductive roles of women, and the power of science would not ordinarily justify interference with procreative liberty....moral, religious, or symbolic concerns that do not have direct, tangible effects on others are not sufficient constitutional grounds for interfering with fundamental rights of persons....[28]

Such arguments are not, however, persuasive if it is granted that the state of knowledge and efficacy associated with IVF is so low as to justify its continued classification as experimental rather than therapeutic.

Moreover, even if one acknowledges a right to procreate or a right to reproductive freedom for every man and woman, it does not follow that the state has no legitimate interest in monitoring or regulating the practice of IVF. The right to freedom with respect to reproduction does not also confer a right on others to withhold or distort information about assisting reproduction. To make autonomous choices, those who seek IVF need accurate information about its efficacy.

Also, acknowledging a negative right—the right to be free from interference with one's sexual and reproductive actions—does not thereby entail the existence of a positive right to be aided or assisted by the state or any other third parties in fulfilling one's procreative desires.[29] The government is under no obligation to assure the availability of IVF to any person who might desire to have an offspring, any more than it is obligated to help find willing mates for the unmarried. Although there may exist a negative right to be free from interference with respect to sexual activities between consenting adults, there is no reason to think that there is a complementary right that entitles anyone to services for the treatment of infertility by any means available, and certainly not by means that are probably still best viewed as experimental, at least at many clinics.

Matters of Respect

If it is true that IVF is properly viewed as experimental, then the issue of the moral acceptability of research aimed at improving the efficacy of the technique needs to be addressed. If those seeking help for infertility require the protection of the state in utilizing procedures that are still experimental, then there is a need to understand the values that should inform regulation in this area.

Low rates of efficacy in combination with a sketchy understanding of why the procedure works are sufficient grounds for classifying IVF as research. A direct benefit of this classification is that it helps answer the question of what purposes ought to govern research in this area.

Individuals seek access to IVF solely for the purpose of creating babies. Some may seek IVF treatment because they are known to be incapable of having a child in any other way. Others may desire IVF because of concerns about the health risks posed to them by a pregnancy. Others may be interested for reasons of convenience, or even to achieve certain eugenic goals.

If IVF is still experimental, it would be hard to justify making it available at the present time to anyone other than those who cannot have children in any other fashion, or for whom a pregnancy is known to be life-threatening. Research ought be undertaken on those whose need for benefits are the greatest, not upon those who seek to be subjects for convenience or for other reasons.

Research, at this point in the evolution of IVF, should concentrate on improving the efficacy of the technique. Basic epidemiological information must be collected. Standardized registries must be created, with mandatory participation of all facilities offering the technique.

Explicit government regulation is not needed to bring this about. All that is necessary is for all third-party payers, both public and private, to insist that every IVF center be enrolled in a national registry if it is to be eligible for reimbursement.

Research must be done upon sperm, ova, and embryos. Work with these human materials is required to ascertain methods by which the technique can be refined so as to produce more babies at less cost.

It is often said that, whatever other moral rules ought govern the treatment of gametes or embryos, these biological materials ought be "treated with respect."[30] One major difficulty with such prescriptions is that it is not self-evident what a commitment to respecting human biological materials entails. Can one espouse a commitment to the respect of embryos while experimenting upon them?

Ironically, if it is true that more research is needed to help advance IVF to the status of a full-fledged therapy, and if research upon techniques aimed at increasing the efficacy of IVF for the infertile ought to have the first priority among other research goals that might be undertaken, then the moral question of the acceptability of research on the unconsenting must still be addressed. If it is true that research that carries risk or danger cannot ethically be done, as an earlier generation of critics of IVF maintained, then it is hard to see how a norm of respect can be compatible with experimentation involving human reproductive materials.

It seems clear that, whatever else is entailed by a norm of respect, it is difficult to insist upon respectfulness solely on the basis of the intrinsic properties possessed by gametes or embryos. Whereas an ova or an embryo is clearly human material, and although there may be the potential present for the development of a human being from such materials, the potential is not inherent solely in gametes or embryos themselves. A woman must be willing to carry these materials in her body under somewhat restricted circumstances for a significant period of time for that potential to be fulfilled.

What we must respect about human reproductive materials is, then, not their potential personhood, but the fact that they are human materials. The attitudes, practices, and interventions we will tolerate with such materials ought to be no more liberal or lackadaisical then those we would tolerate where other human materials, such as organs or tissues, are concerned.

To show respect may mean prohibiting sperm, ova, or embryos to be bought and sold. It may also mean prohibiting the creation of an embryo solely to experiment upon it. To intentionally create embryos to experiment upon is to treat human life solely as a means and not as an end.

A commitment to respect may also entail that certain routines and customs ought to govern the procurement and manipulation of such materials. There may be licensure or registration requirements imposed on those working with these materials. Those who wish to utilize such materials must obtain the permission of those from whose bodies they came, and from committees or other groups that society may deem appropriate, to ensure that the goals undertaken and the benefits to be produced by using such materials are consistent with public sensibilities.

There is a long history in Western law and morality that establishes the nonproperty status of persons and their bodies as a sign of special respect for what is human. It would seem imperative, in order to remain consistent with this tradition, to extend the same nonproperty, noncommercial status to human reproductive materials. This is not done from an economic motive, but from a moral motive. Placing human materials outside the framework of the market is to accord them special status and, thus, to respect them. To be meaningful, such a restriction would have to apply to both the sources of human materials and healthcare providers who wish to obtain them.

Those seeking IVF are at a special disadvantage, in that they are especially vulnerable to the requests of researchers for consent to a form of intervention that they desperately seek. It may make sense to insist upon independent committee review for each and every case of procurement of human reproductive materials.

Consent is not the sole determinant of the acceptability of research on human reproductive materials. It might be argued that the potentiality inherent in such materials forces the question upon us of whether research that might involve damage to, or the destruction of, such materials is morally licit.

If the aim of such research is to improve the efficacy of IVF, then it would seem possible to justify the use and even the destruction of these materials. The reason is quite simple—unless such research is done, the potential inherent in human reproductive materials is likely to remain only that—potential. If the goal of allowing the potential of a sperm, ova, or embryo to be expressed in the form of a baby is a morally justifiable one, then it would seem morally acceptable to allow the routine fulfillment of that goal.

Those who wish to see IVF accepted as a routine element of medical practice have thought it best to use the language of therapy in describing the technique. However, although it is true that IVF seems to entail no harm for those children who are created in this manner, it is also true that medical science is not yet capable of making the technique work in a reliable way without imposing significant burdens on those involved. The only way to modify this situation is to come to terms with the moral reality that it is only by encouraging further research on human reproductive materials that respect for human life both present and potential can be demonstrated.

References

[1]Fishel, S. (1986) IVF-Historical Perspective, in *In Vitro Fertilization* (Fishel, S. and Symonds, E., eds.), IRL Press, Oxford, pp. 1–16.

[2]Edwards, R., Steptoe, P. and Purdy, M. (1970) Fertilization and cleavage in vitro of preovulator human oocytes, *Nature* **227,** 1307–1309.

[3]Ramsey, P. (1972) Shall we reproduce? *JAMA* **220,** 1346–1350.

[4]Kass, L. (1971) Babies by means of in vitro fertilization: Unethical experiments on the unborn? *N. Eng. J. Med.* **285,** 1174–1179.

[5]Ramsey, P.

[6]McCormick, R. (1974) Proxy consent in the experimentation situation, *Perspec. In Biol. and Med.* **18,** 2–20.

[7]Tiefel, H. (1982) Human in vitro fertilization: A conservative view, *JAMA* **243,** 3235–3242.

[8]Congregation for the Doctrine of the Faith (February 22, 1987) Instruction on respect for human life in its origin and on the dignity of procreation (Congregation, Vatican City).

[9]Kass, L. (1979) Making babies revisited, *The Public Interest* **54,** 32–60.

[10]Caplan, A. (1985) Ethical issues raised by research involving xenografts, *JAMA* **254,** 3339–3343.

[11]Katz, J. (1985) *The Silent World of Doctor and Patient* (Free Press, New York).

[12]Caplan, A. (1988) Human experimentation and medical technology, *IEEE Journal* **7,** 74–76.

[13]US Congress, Office of Technology Assessment (1988) *Infertility: Medical And Social Choices* (GPO, Washington, DC).

[14]Levine, R., "On the Relevance of Ethical Principles and Guidelines Developed for Research to Health Services Conducted or Supported by the Secretary," *Ethical Principles and Guidelines for the Protection of Human Subjects of Research* Appendix I:1-91, DHEW Publication No. OS 78-0011 (Washington, DC, 1978).

[15]Department of Health and Human Services, "Rules and Regulations," 45 CFR 46, 46. 102 (e), (March 8, 1983).

[16]American Fertility Society Ethics Committtee (1986) Ethical considerations of the new reproductive technologies, *Fertil. and Steril.* **46** Suppl. 1, 1s–94s.

[17]Medical Research Council/Royal College of Obstetrics, Second Report, Voluntary Licensing Authority for Human IVF and Embryology (1987).

[18]US Congress, Office of Technology Assessment.

[19]Caplan, A., "Ethical and Policy Considerations Regarding In Vitro Fertilization," Testimony to Subcommittee on Regulation and Business Opportunities, US Congress (June 1, 1988).

[20]Medical Research Council/ Royal College of Obstetrics.

[21]American Fertility Society Special Interest Group (1988) In vitro fertilization/ embryo transfer in the United States: 1985 and 1986 results from the National IVF/ET Registry, *Fertil. Ster.* **49,** 212–215.

[22]Edwards, K. (1988) Reproductive technology, *Ohio Medicine* **44,** 183–193.

[23]Holmes, H., Hoskins, B., and Gross, M. (eds.) (1981) *The Custom Made Child?* (Humana,Clifton, NJ).

[24]Rowland, R. (1987) Making women visible in the embryo experimentation debate, *Bioethics* **1,** 179–188.

[25]Jones, H. (1986) *In Vitro Fertilization* (Williams and Wilkins, Norfolk, Baltimore).

[26]American Fertility Society, Ethics Committee.

[27]American Fertility Society, Ethics Committee.

[28]American Fertility Society, Ethics Committtee.

[29]H. Shue (1980) *Basic Rights* (Princeton University Press, Princeton, NJ).

[30]US Congress, Office of Technology Assessment.

Surrogate Motherhood

Surrogacy Arrangements

An Overview

Dianne M. Bartels

A surrogacy arrangement is a social mechanism developed to address the needs of a couple who are unable to produce a pregnancy or of a woman who is unable to carry pregnancy to term. With the use of artificial insemination and in vitro fertilization, a couple now can make an arrangement with a woman who is willing to conceive a child, carry it to term, and then surrender parental rights. That woman is called the "surrogate" or "contract" mother. In standard surrogacy arrangements, the sperm of the contracting father is utilized. This provides the advantage of producing a genetically related child.

Although formal surrogacy arrangements have existed since 1976, they first attained national visibility when Mary Beth Whitehead, the surrogate mother in the "Baby M" case, refused to honor the contract and insisted on keeping her baby. This case was addressed in the New Jersey courts in 1987. Legislators across the country have since proposed bills to regulate, permit, or prohibit surrogacy arrangements. Prior to consideration of these public policy initiatives, it is important to understand the context in which surrogacy occurs: who participates, for what reasons, and at what cost. These dimensions are the focus of this chapter.

The Problem of Infertility

Infertility, defined as the inability to produce a pregnancy after 12 months of unprotected intercourse, affects an estimated 2.4 million couples in the US today and is the major reason couples seek surrogacy arrangements.[1] Although the actual rate of infertility has not increased since 1965, the services available to address the causes have changed significantly. Treatment of infertility parallels other areas of medical practice, in which developing technologies and pharmacological innovations provide new alternatives to correct medical and social problems.

We have learned that we can explore space, break genetic codes, and replace what does not work, even if it is a human heart or a set of lungs. Thus, as a society we have come to believe that we need not accept the scourges of disease or disability, including the "handicap" or "disease" of infertility. Despite progress in the treatment of infertility, the success rate is still low.[2] When medical and surgical interventions fail, couples may look to artificial insemination, adoption, in vitro fertilization, or surrogacy. Compounding the frustration of childlessness is the fact that individuals now spend years on waiting lists for adoption or are told that they are ineligible because of the ages at which they have decided to create a family.

Types of Surrogacy Arrangements

The William Stern–Mary Beth Whitehead (Baby M) arrangement is a typical surrogacy-for-pay arrangement.[3]

In altruistic surrogacy, on the other hand, a couple makes an arrangement with a close friend or relative without the involvement of attorneys or brokers or the payment of fees. It also is possible for the couple to supply the egg and sperm to produce a child genetically related to both of them and to have the embryo implanted in the womb of a woman able to carry it until delivery. Additional possibilities have been created by technological advances. One can now donate ova or sperm and preserve them, by freezing, for use at a later time. These advances open the possibility of parenthood to whomever has the resources to contract for it, and raises the concern that we could begin to produce "designer children."

A child *could* be produced through the efforts of five different contributors to the babymaking process—the "genetic" parents who contribute the egg and sperm, the "social" parents who contract to raise the resulting child, and the "surrogate" who provides the gestational services for the embryo implanted in her womb[4]—thereby separating genetic, gestational, and social parenting. It is this separation of the elements of parenting that raises moral questions about the reproductive process itself, the moral status of the child, and the rights and responsibilities of each of the participants.

Demographics of Surrogacy

The earliest legal contracts for surrogacy arrangements were initiated by Noel Keane, the broker in the Baby M case, who helped arrange eight surrogate births between 1975 and 1981.[5] There were over 600 children in the United States known to have been born through surrogate arrangements by May of 1988.[6] Seventeen recognized centers in the United States provide brokering and legal services. Only some of these centers comprehensively assess the physical and emotional status of the parties to the contract. Center directors and staff are diverse, and include physicians, psychologists, counselors, and attorneys.[7]

For the couple who wants a child, the cost of participation averages $20,000–35,000. These couples come from all socioeconomic classes and cultural backgrounds. They are as likely to be blue collar as white collar. Some have taken second and third mortgages to attain resources for these contracts. The usual fee for a surrogate is $10,000, paid following the delivery of the child and the relinquishment of parental rights. The majority of surrogates surveyed are married, have at least one child, and have an average family income of $25,000.[8] Hilary Hanafin's survey of 89 women who have been surrogates indicates a mean age of 28 years and a mean of 2.0 children. "A ratio of five to one of the women were Anglo."[9]

An estimated 20,000–35,000 women contacted surrogate-parenting programs between 1980 and 1987 to inquire about becoming surrogate mothers. "Clinic directors reportedly accept twice the number of surrogates as couples to ensure availability."[10]

Reasons for Participation in Surrogacy Arrangements

Why do some couples stop short of surrogacy in their attempts to have a child and others do not? What are the characteristics of the women who contract to provide the services? Why would one want to be involved in facilitating these arrangements? Since the practice is new and survey samples small, the data available about the participants in surrogacy arrangements is sketchy at best.

Who are the couples who become involved in this tedious, risky, and costly process? In addition to paying the fee, the couple "must traverse a legal minefield" and deal with the sometimes negative reactions of family members and friends.[11] One author notes, "What is different about these couples—what sets them apart from other couples who happen to be infertile is a driving need to have a child."[12] For many of these couples, "money is no object" in pursuit of this primary goal. Frustration with adoption processes and the wish to have a genetically related child also contribute to their desire to exert some control over the process of creating their child.

What is unique about the women who provide the "service" of surrogacy? Enjoyment of the pregnant state and the desire for another pregnancy were sited by a majority of surrogates interviewed. In Hanafin's survey, three participants had adoption in their history, more than a third had abortion in their history, and 13% had had more than one abortion. For some, their unresolved feelings related to these earlier reproductive experiences were a factor in the desire to become a surrogate. They also expressed empathy for childless couples and the "desire to do something remarkable with their lives and to make a unique contribution."[13] Although altruism is consistently defined as a primary motivating factor, few have yet been identified who said they would perform this service without payment.[14,15]

Why would someone external to the situation want to become involved in these arrangements? One clinic director notes that her own inability to conceive and empathy for couples who share her plight are reasons for providing the surrogacy contract.[16] Surrogacy is also a new business with growth potential. After "Sixty Minutes" highlighted

Noel Keane and his business of brokering surrogacy arrangements, his center experienced a major increase in the number of inquiries from prospective clients. Revenues generated from the couples who wish to become parents average $20,000–35,000, with some reports of fees up to $45,000 per child.

Directors of surrogacy centers are not the only participants who stand to gain financially from such arrangements. They create a potential new revenue source for hospitals. "The market for fertility services is increasingly attractive, with more couples in their 30s with more discretionary income trying to conceive at a time in their lives when fertility is naturally lower."[17] Regulations could increase or decrease the opportunities to provide service and generate fees. Legislation that makes infertility a "precondition" to a contract for surrogacy services will expand the clientele for infertility testing centers. Requirements for physical and psychological assessments and genetic testing also will promote expansion of these services. Alternatively, legislation that bans or sets major limits on surrogacy arrangements would be an impediment to development of these new business opportunities.

Public Policy Alternatives

The description of the participants in surrogacy arrangements indicate some of the ethical concerns that may be addressed by legislation. The lack of any requirement for assessment and counseling *prior* to initiation of the contracts raises a question about the physical or emotional "fitness" of the would-be surrogates and the social parents. The profit potential coupled with the lack of policies to regulate either the agencies or the credentials of the people who manage these centers means that virtually anyone can "hang out a shingle" and open a new business. The "desperation" of the couples seeking children and the economic disparity between the contracting couple and the surrogate mean that all of the participants are potentially subject to coercion or exploitation. Legislators in Minnesota and other states became concerned when it was noted that Noel Keane was advertising for potential surrogates in college newspapers. Moral concerns about the "commodification" of children as a means to another's (the parents') ends and concerns about the undermining of respect for women and family are additional reasons for the initiation of legislation on surrogacy.

Interestingly, none of the health care professional associations has taken a position regarding surrogacy, nor have they proposed structures for credentialing or regulating clinics or practitioners. The American Fertility Society merely advises the individual physician to be cautious in his or her involvement in surrogacy arrangements.

State legislators, on the other hand, have become very involved in developing responses. Bills addressing surrogacy have been proposed in states across the US.[18] Four major types of responses have been proposed. One is to ban or to criminalize the practice of surrogacy and the parties to it, with the harshest penalties for brokers and marketers, who have the prospect of the greatest financial gain. A second is to regulate surrogacy, to afford protection to both the parties involved in the contract and the potential child. Protections include such things as providing for health coverage for the surrogate or support for the child who may not be seen as acceptable to the contracting parents. Elimination of genetic testing and abortion requirements from the contract is another protection that is commonly proposed.[19,20] A third alternative encourages the practice of surrogacy by removing it from the "baby selling" statutes, as has occurred in Nevada. A fourth response is to consciously ignore the issues and allow the future to be determined by the courts as disputes arise.

Legislative and judicial responses address the moral concerns of those proposing them. How one weighs the potential benefits and harms, as well as the view of the role of the state in reproductive decisions, will determine the position one takes in relation to surrogacy arrangements. Those who oppose regulation of the practice of surrogacy generally base their opposition on two premises. One is the fundamental right to privacy, including noninterference in reproductive decisions as provided in *Roe v. Wade*. This position asserts that a woman has the right to determine what she will do with her body, even if that means doing something that others would see as immoral or demeaning. Whether the outcome of her decision is positive or negative, it is her choice.

A second argument opposing regulation is an autonomy-oriented position holding that the state should not interfere with contracts of consenting adults. It asserts that individual actions should be controlled only when they are a danger to the community or to the common good.

Subsequent chapters pose alternative views about the practice of surrogacy. They also identify policy options and unanswered questions related to the status of embryos, fetuses, children, women, men, marriage, and the family that are central moral considerations any time one addresses change in the practice or regulation of human reproduction.

References

[1]Ellis, G. (1989) Infertility and the Role of the Federal Government, this volume.

[2]Ellis, G.
[3]Merrick, J. (1989) The Case of Baby M, this volume.
[4]Tagatz, G. E., Medical Techniques for Assisted Reproduction, this volume.
[5]Gelman, D., and Shapiro, E. Infertility: Babies by contract, *Newsweek* **106,** 19, (November 4, 1985).
[6]Ellis, G.
[7]Overvold, A. Z. (1988) *Surrogate Parenting* (Pharos Books, New York).
[8]Overvold, A. Z.
[9]Hanafin, H., "Surrogate Parenting: Reassessing Human Bonding," presented at APA Convention, New York, New York (August 28, 1987).
[10]Overvold, A. Z.
[11]Gelman, D., and Shapiro, E. Infertility: Babies by contract, *Newsweek* **106,** 19, (November 4, 1985).
[12]Overvold, A. Z.
[13]Hanafin, H.
[14]Ellis, G.
[15]Overvold, A. Z.
[16]Overvold, A. Z.
[17]Kelly, Mary Ann (1987) "Meditrends: In Vitro Fertilization Services," *A. H. A. Hospital Technology Series* (December, 1987).
[18]Merrick, J.
[19]Bopp, J. (1989) Surrogate motherhood agreements: The Risks to Innocent Human Life, this volume.
[20]Merrick, J.

The Case of Baby M

Janna C. Merrick

Many changes have taken place in recent years in the area of assisted reproduction. One of the more controversial is surrogate motherhood, which is used primarily when a woman either is infertile or, for other reasons, cannot or will not bear her own biological child. Her partner's sperm is artificially inseminated in a "surrogate," who agrees to surrender the child to the couple at birth. Such surrogates are normally paid.

For childless couples who desperately seek to become parents, surrogacy may seem like a viable option. However, as the Baby M case shows, such an option may be filled with risk. The well-publicized struggle between Baby M's biological parents has raised concern about the legal implications of such an arrangement. This chapter will trace the development of the case, examine the findings of the New Jersey courts, and conclude with a discussion of the legal ramifications of the dispute.

Background Facts

The story of Baby M received worldwide attention. Forty-year-old William Stern, a biochemist, and his wife Elizabeth, a pediatrician, contacted the Infertility Center of New York (hereinafter referred to as ICNY), which specializes in surrogate parenting, ovum transfer, and in vitro fertilization with implantation in surrogates. ICNY operates as a middleman, matching contracting couples with possible surrogates. It is run by attorney Noel Keane, who also has agencies in Michigan and Nevada, and an informal operation in the Netherlands.*

Mrs. Stern had a mild form of multiple sclerosis and believed that pregnancy would threaten her health. Mr. Stern felt strongly about having a biologically related child, since both his parents were dead, and other members of his family had been killed in the Holocaust. He had no known living relatives. The couple was middle class, he with a doctorate and she with both a doctorate and an MD They reported their income in the $90,000 range.

ICNY matched the Sterns with Mary Beth Whitehead, a 29-year-old homemaker whose personal life differed sharply from theirs. She left high school at age 15, married shortly thereafter, and had two children. The family moved frequently, often living with relatives. In 1978, she and her husband, Richard, separated briefly, and she received public assistance. Later, they filed for bankruptcy, and during the custody battle they were fighting foreclosure on their home. Richard Whitehead was a sanitation worker and supported the family on $28,000 per year. He had an unsteady employment record and admitted to unresolved alcohol problems.

When the Sterns and Whiteheads met, Mary Beth Whitehead indicated eagerness to provide the service. She asked for an annual letter and photo of the child, and assured the Sterns they would not find her on their door-

---

*Keane initially opened an office in Frankfurt, West Germany, to refer European couples to his American agencies. The West German government closed that office, and now his operation is run informally from the Netherlands.

step. As William Stern was later to say, "It seemed too good to be true." He and the Whiteheads signed the surrogacy agreement on Feburary 6, 1985. Elizabeth Stern was not a party to it. The contract described her as being infertile, but did not mention her multiple sclerosis (*see* Appendix A).

The contract provided by ICNY was typical of other surrogacy arrangements and was weighted in favor of the contracting father. For the most part, it obligated the Whiteheads to perform certain duties, and guaranteed to Stern certain rights. Mary Beth Whitehead would undergo artificial insemination and freely surrender custody to Stern immediately after delivery. If he died before the birth, custody would be surrendered to Elizabeth Stern. The Whiteheads would not form any parent–child bonds, meaning they would not exercise powers of a parent, or seek custody or visitation. They would terminate their parental rights, thereby allowing Elizabeth Stern to adopt the child. Mary Beth Whitehead would submit to prenatal testing, and if it was determined that the child had defects, would undergo an abortion if demanded by William Stern. Once the child was born, William Stern would assume legal responsibility (such as financial support) even if there were defects or handicaps. The contract did not say, however, that he would accept custody under those circumstances.

Mary Beth Whitehead was to receive $10,000 for her "services." If she miscarried prior to the fifth month of pregnancy, there would be no compensation, and if she miscarried afterwards or the baby was stillborn, her compensation would be $1000. In addition to the fee, Stern would pay any medical expenses not covered by her health insurance. In a separate contract, he agreed to pay ICNY a nonrefundable fee of $7500 for assisting him in the selection of a surrogate mother. The agency would provide legal representation in his negotiations with her if he so desired. Also as part of this contract, Stern cited that he was free from disease and hereditary medical problems.

The Whiteheads did obtain some legal advice prior to signing the surrogacy agreement with Stern. ICNY had previously considered Mary Beth Whitehead as a possible surrogate for another couple. The Whiteheads saw an attorney who was referred to them by ICNY.[1] He explained the contract to them and suggested some marginal changes.

Stern and the Whiteheads agreed to the contract, the bargain was struck, and artificial insemination proved successful. Initially, relations between the Sterns and Whiteheads were cordial. They had dinner together, talked on the telephone, and the Whiteheads' daughter spent the night at the Sterns' home. The pregnancy progressed without complication, and Baby M was born March 27, 1986. Her biological father named her Melissa Stern, and her biological mother named her Sara Whitehead. Mary Beth Whitehead had changed her mind and did not want to give up her baby.

When the Sterns arrived at the Whitehead home several days later to claim the child, Mrs. Whitehead was in a state of hysteria, claiming she had made a terrible mistake and could not go through with the bargain. She did, however, allow the Sterns to leave with the child. Shortly thereafter, she met with the Sterns at their home and begged to take the child for a week. They consented, and she later notified them of her intention to keep the child and to leave the country if necessary.

The Sterns obtained a court order demanding temporary custody from Superior Court Judge Harvey Sorkow and appeared with the police at the Whitehead home on

May 5th. In the confusion, the baby was handed out a back window to Richard Whitehead, who spent the night with relatives. The next day, the family fled to Florida and disappeared for 87 days. The Whiteheads left their older children with grandparents and were on the run, staying in about 20 different locations. During this time, Whitehead telephoned William Stern, threatening to kill herself and the baby, and indicated that she would falsely accuse him of sexually molesting her older daughter. The Sterns hired detectives, obtained a Florida court order, and with the help of local authorities returned the child to New Jersey, where they were granted temporary custody in July.

On August 13th, Judge Sorkow appointed an attorney as guardian *ad litem* to represent the interests of the child, because he felt that the Sterns and the Whiteheads had personal agendas, and that Baby M needed independent counsel to represent her. He also permitted Mrs. Whitehead supervised visitation with the child for four hours per week. The actual trial began on January 5, 1987. The case would have been held in closed court to protect the interests of the child; however, Mary Beth Whitehead's mother contacted the news media, and major press coverage began.[†]

The Trial Court Decision

The trial, presided over by Sorkow, was conducted in two stages. The first dealt with the legality of the contract, and the second considered the custody arrangements.[2]

[†]Information on the facts of the case prior to the trial is based on accounts in *The New York Times,* as well as the procedural histories provided in the Sorkow and Wilentz decisions.

Judge Sorkow would have to be a pathbreaker. There were no statutes or prior court decisions in New Jersey on which to rely. The entire trial took six weeks, and 38 witnesses were called. Both sides called on experts to testify to the relative emotional stability of the Sterns and the Whiteheads. The guardian *ad litem* also called on such experts. The decision was handed down on March 31, 1987.

The court found that the contract in its essentials was valid, enforceable, and not subject to the adoption laws of New Jersey. Whitehead argued unsuccessfully that the Sterns and ICNY committed fraud during the negotiation of the contract, because the contract specified that Mrs. Stern was infertile when, in fact, she had never been tested for fertility. Second, Mrs. Stern's multiple sclerosis was not revealed to Whitehead, which she argued would have deterred her from serving as their surrogate. Finally, ICNY did not reveal to her, or to the Sterns, that the psychological tests performed by the agency, prior to Whitehead becoming a surrogate, indicated that she would have difficulty surrendering her infant.

Whitehead also argued that the terms of the contract were oppressive and should not be enforced. However, Judge Sorkow found that she had previously been pregnant and knew the risks of bearing a child. He pointed out that Stern, too, incurred a risk, because he would have parental responsibility for the child even if born with abnormalities. The court found that Whitehead knew what she was bargaining for, had received some legal advice, and then reneged on her promise to surrender the infant. Based on the 1973 US Supreme Court decision in *Roe v. Wade*,[3] Sorkow did strike down the clause in the contract that would have required abortion upon demand of the biological father.

Judge Sorkow further upheld the contract by drawing an analogy between surrogate motherhood and selling sperm. In essence, he argued, a man becomes a "surrogate father" when he sells his sperm. Equal protection guaranteed by the Constitution then requires that a woman also be allowed to sell her procreative abilities. Thus commercial surrogacy was legal, the contract between William Stern and the Whiteheads was enforceable, and Mary Beth Whitehead breached it when she refused to surrender the child and failed to renounce her parental rights.

The court then turned to the custody issues because, regardless of the enforceability of the contract, the best interest of the child was still the overriding factor in deciding who would raise her. In this regard, Judge Sorkow was highly critical of Whitehead's behavior. She had been less than truthful on the witness stand about a number of issues, including sworn testimony that she believed the child was biologically her husband's, when she knew that he had had a vasectomy nine years before. Such dishonesty led the judge to believe that she would not be truthful with the child about her origins, which would become important as she grew older. Judge Sorkow noted many of the problems discussed above, including Whitehead's failure to complete high school, her early marriage, the family's financial instability, Richard Whitehead's alcohol problems, and the Whiteheads' failure to honor a court order and subsequent flight from the state.

He did not claim that Mary Beth Whitehead was an unfit mother. To the contrary, he indicated that she had been a good mother to her other two children. However, her behavior during the custody crisis raised doubts about her emotional stability. Moreover, the home life that she could provide would not be as beneficial to the child as that which the Sterns could provide. She was described as being manipulative, impulsive, and exploitive, whereas the Sterns were described as loving, nurturing, and willing to foster the child's independence. Witnesses called by the guardian *ad litem* testified that Whitehead would have trouble subordinating her own needs to those of her child.

Thus, Judge Sorkow ordered that the surrogacy agreement be enforced, with Mr. Stern having custody. Whitehead's parental rights were terminated, and Mrs. Stern was allowed to adopt the child. The birth certificate was amended to show that her name was Melissa Stern, and the Whiteheads, including their friends and relatives, were restrained from interfering in the parenting rights of the Sterns. Finally, Whitehead was to receive the $10,000 surrogacy fee.

The New Jersey Supreme Court Decision

The Whiteheads appealed to the New Jersey Supreme Court, which heard arguments on September 14, 1987. In an unanimous decision written by Chief Justice Robert N. Wilentz, and released on February 3, 1988, the justices overturned the lower court decision that paid surrogacy agreements were legal. They affirmed, however, that it was in the best interest of the child for William Stern and his wife to retain custody. Mary Beth Whitehead was recognized as the legal mother and was referred to throughout the decision as "mother," rather than "surrogate mother." She was awarded visitation, the extent of which was to be decided by a lower court. The adoption by Elizabeth Stern was voided.[4]

In setting aside the contract, the Court found that it conflicted with New Jersey law prohibiting the use of money in connection with adoptions. Clearly, the fees paid

Implications for Surrogacy

to Whitehead were for the adoption and not for her personal services. If the child were stillborn, the fee was only $1000, yet a mother provides the same "service" whether she delivers a healthy child or a stillborn infant. According to the Court,

> Mr. Stern knew he was paying for the adoption of the child; Mrs. Whitehead knew she was accepting money so that a child might be adopted; the Infertility Center knew that it was being paid for assisting in the adoption of a child.[5]

This is clearly baby-selling, and is illegal, and perhaps criminal. It is

> the sale of a child, or, at the very least, the sale of a mother's right to her child, the only mitigating factor being that one of the purchasers is the father. Almost every evil that prompted the prohibition on the payment of money in connection with adoptions exists here.[6]

The Court stated quite bluntly that, in a civilized society, there are things money simply cannot buy.

The Court then turned to the issue of parental rights, which allow mothers and fathers to participate in the rearing of their children. New Jersey law requires proof of parental unfitness or abandonment before terminating such rights. The Court addressed Whitehead's role as a mother, pointing out that it was never even alleged that she was unfit. The lower court had, indeed, said that she was a good parent for her other two children. Clearly she had not abandoned her baby, and thus her parental rights could not be terminated. Moreover, New Jersey adoption laws allow a time period for a mother to change her decision in a private-placement adoption. Whitehead made it known to the Sterns almost immediately after the birth that she did not want to give up her child, and she was well within the time period allowed.

The Court also found that the contract violated established New Jersey public policy on child custody, because it allowed the natural parents to decide prior to birth who would raise the child. Custody is based on the best interest of the child and cannot be determined prior to birth. The contract also violated the principle that children should remain with and be raised by both natural parents, and should not be permanently separated from either.

Having set aside the contract, the Court turned to the issue of custody. It analyzed the behavior of the Sterns and Whiteheads during the dispute and described the type of home environment each could provide. It reached the same conclusion as that reached by the lower court. The Sterns were depicted as stable, loving, and nurturing, having provided a good home for the child during the year and a half she had been with them. Mary Beth Whitehead was viewed as unstable, at least during this crisis. She had told inconsistent stories and was unable to separate her own needs from those of her children. The Court also noted her apparent contempt for professional psychiatric help, which it felt was needed in this situation. None of the expert witnesses, even those testifying in her behalf, had stated or implied that she should have custody.

Most importantly, the three experts hired by the baby's guardian *ad litem*, who the Court argued were free of bias, recommended that the Sterns have custody. Shortly after the Supreme Court appeal began, Mary Beth Whitehead announced that she was pregnant by another man, divorced her husband Richard, and quickly remarried. Noting this, the guardian *ad litem* recommended that visitation between Whitehead and her daughter be postponed until the child reached maturity. The Supreme

Implications for Surrogacy

Court argued that this would be highly unusual and ordered visitation, the extent of which was to be determined by a lower court. The Court also ordered that a judge other than Sorkow be assigned to the case.

The case was remanded back to Superior Court, which granted Whitehead broad rights to see her daughter. The Superior Court also ordered both Whitehead and the Sterns to participate in mental health counseling, and prohibited both families from publicizing the child's activities or selling movie rights without court approval.[7]

Although the New Jersey Supreme Court was not complimentary toward Whitehead, it criticized Judge Sorkow for his harshness toward her. It was unrealistic to expect her to surrender her child without a struggle.

> We do not know of, and cannot conceive of, any other case where a perfectly fit mother was expected to surrender her newly born infant, perhaps forever, and was then told she was a bad mother because she did not. We know of no authority suggesting that the moral quality of her act in those circumstances should be judged by referring to a contract made before she became pregnant.[8]

This is a case in which both sides suffered, and in which most likely there will be suffering in the future. According to the New Jersey Supreme Court, it comes from the evils of paid surrogacy.

> In the scheme contemplated by the surrogacy contract in this case, a middle man propelled by profit, promotes the sale. Whatever idealism may have motivated any of the participants, the profit motive predominates, permeates, and ultimately governs the transaction. The demand for children is great and the supply small.[9]

Legal Implications of Baby M

So where does the Baby M case leave us? The findings of the New Jersey Supreme Court apply only in that state, and could be modified or reversed if the state legislature acts. It should be noted, however, that the New Jersey Supreme Court is considered a leader both in family law and in medical ethics, and thus the decision might provide a model for other courts. We can say clearly that, unless the legislature acts, paid surrogacy violates New Jersey's adoption and baby-selling laws, and is therefore prohibited. Unpaid surrogacy remains legal, but falls under the state's laws on private-placement adoption, which allow the biological mother to change her mind within a reasonable time period. It is highly unlikely that unpaid surrogacy will flourish.

William Stern and Mary Beth Whitehead are Melissa's legal parents, and both have rights to participate in her rearing and responsibilities toward her care. Neither parent can be permanently separated from her unless unfitness or abandonment are proved. Physical custody has been, and will continue to be, determined in accordance with the best interests of the child.

William Stern and Mary Beth Whitehead are clearly not "done" with one another because they are permanently connected through Melissa. New Jersey allows custody cases to be reopened when there are changed circumstances that may affect the welfare of the child. Therefore, it is possible that Melissa's parents will oppose one another in court over changes in custody, visitation, and support. At some point, Melissa will have the opportunity to choose for herself where to live. A noncustodial parent is normally required to pay child support to the custodial parent. Should Mary Beth Whitehead become financially stable—a real possibility in the event of book or movie rights—Stern could sue her for support. Should custody be shifted to Whitehead, she could sue Stern for support. They could contest one another over a number of other issues, including Melissa's medical care, education, and religious training.

Most of the other actors in the dispute have few, if any, rights. Elizabeth Stern, who took a part-time leave of absence from her medical school teaching position to raise Melissa and whom the child probably identifies as her mother, has no legal claim to participate in decisions about her life. Should William Stern die, custody could shift to Whitehead, who is the child's legal mother. If the Sterns divorce, William would have an advantage over Elizabeth in seeking custody, or to make the situation more complicated, Whitehead could argue that this constituted a changed circumstance and could mount a custody battle herself. Richard Whitehead, who supported his wife through the pregnancy, gave up his job to flee with her to Florida, and stood by her during the legal battle, has no claim. The Whitehead's divorce settlement does, however, specify that he will receive a share of any book or movie royalties, as well as a share of the out-of-court settlement that ICNY paid to them in a separate legal action. Dean Gould, Mary Beth Whitehead's second husband, has no legal rights regarding Melissa's care or custody.

This case raises a number of interesting questions. What does it say regarding socioeconomic class exploitation? Should members of one socioeconomic class be allowed to buy the procreative abilities of women from another class, and if so, for how much? Although the $10,000 fee that Whitehead was to receive seems generous, an *amicus* brief filed by the Rutgers Women's Litigation Clinic in the Baby M case pointed out that payment for her "services" amounted to $1.57 per hour.[10] A more fundamental question is: should those with wealth be allowed to buy the children that result from such a union?

From the outset, the Sterns had certain advantages. They were middle class, highly educated, and knew how to use the legal system. The surrogacy agreement favored their interests. They had the resources needed to mount a custody battle. It is estimated that they spent $60,000–70,000 on legal fees, detectives, and expert witnesses.[11] The costs of the state Supreme Court appeal were additional. Their greater wealth could not only provide them with some of the best family law attorneys in New Jersey, but also provide a more lucrative life style for the child, an issue that both courts took into serious consideration in awarding them custody. Mary Beth Whitehead was a high school dropout who wanted the $10,000 to help pay the educational expenses of her other two children. Her attorney was far less experienced in family law, and she could not begin to compete with the Sterns in the type of life style she could provide. The PhD and MD easily won custody over the high school dropout.

There are other issues as well. What does this case mean in terms of mother's rights vs father's rights? Although a contracting father generally has more income than the mother, he faces other liabilities. He would not spend close to $20,000 on a surrogacy arrangement if he did not seriously want a biologically related child. Yet the bias in the family law system makes it difficult for fathers to gain custody, particularly of a newborn. The New Jersey Supreme Court noted this in its ruling.

> When a father and mother are separated and disagree, at birth, on custody, only in an extreme, truly rare, case should the child be taken from its mother...before the dispute is finally determined by the

court on its merits. The probable bond between mother and child, and the child's need, not just the mother's, to strengthen that bond, along with the likelihood, in most cases, of a significantly lesser, if any, bond with the father—all counsel against temporary custody in the father. A substantial showing that the mother's continued custody would threaten the child's health or welfare would seem to be required.[12]

The Court noted that perhaps Judge Sorkow erred in awarding temporary custody to the Sterns almost immediately after the birth. However, regardless of such error, the child seemed happy and well adjusted, and the Court did not want to disrupt her life by shifting custody to her mother. It is interesting to speculate whether there would have been a different outcome had Mary Beth Whitehead been college-educated, middle class, and with a stable family situation. It is possible that a father could enter a surrogacy agreement, have the mother change her mind, and then see her win both custody and child support.

New Jersey is not the only state faced with surrogacy disputes, and some are following different routes. Baby-selling is illegal in all states, and about half prohibit payment in connection with adoption.[13] However, not everyone agrees that paid surrogacy violates these laws, which were enacted to protect vulnerable, usually unwed, mothers from making hasty decisions about surrendering their children. Thus, in 1986, the Kentucky Supreme Court found that paid surrogacy did not violate its baby-selling statute if agreements were made prior to conception and the mother was allowed to change her mind before relinquishing her parental rights. The Court argued that

the essential considerations for the surrogate mother when she agrees to surrogate parenting procedures are *not* avoiding the consequences of an unwanted pregnancy for fear of the financial burden of child rearing. On the contrary, the essential consideration is to assist a person or couple who desperately want a child but are unable to conceive one in the customary manner, to achieve a biologically related offspring.[14]

State legislatures have also begun to act. In the aftermath of Baby M, 73 bills were introduced in 27 states. Passage has been rare. Louisiana and Michigan prohibited paid surrogacy, whereas Nevada exempted such arrangements from laws that ban payment for adoption. Arkansas provides that, if the surrogate is unmarried, the child shall, for legal purposes, be that of the woman intended to be the mother by contract. Most bills pending in 1987 would have allowed surrogacy in some form, although not necessarily commercial surrogacy.

Conclusion

Perhaps surrogacy remains a reasonable alternative for childless couples, but it is not risk free. Public policy is sorely lacking, and the legal system is unprepared to resolve disputes that invariably will arise as fathers and "surrogate" mothers contest one another over their children. In the case of Baby M, her parents differed in terms of life styles and values. As a result of their ordeal, they are bitter and angry at one another. However, William Stern is the father, and Mary Beth Whitehead is the mother. Both have rights and responsibilities for their child's care. These two people, who could not even agree on their child's name, must now share in raising her.

References

[1]*In the Matter of Baby M* 537 A.2d 1227, 1247 (N.J. 1988).
[2]*In RE Baby M* 525 A.2d 1128 (N.J. Super. Ch. 1987).
[3]*Roe v. Wade*, 410 U.S. 113, 93 S. Ct. 705, 35 L.Ed. 2d 147 (1973).
[4]*In the Matter of Baby M* 537 A.2d 1227 (N.J. 1988).
[5]*In the Matter of Baby M* 537 A.2d 1227, 1241 (N.J. 1988).
[6]*In the Matter of Baby M* 537 A.2d 1227, 1248 (N.J. 1988).
[7]225 N.J. Super. 267 (1988).
[8]*In the Matter of Baby M* 537 A.2d 1227, 1259 (N.J. 1988).
[9]*In the Matter of Baby M* 537 A.2d 1227, 1249 (N.J. 1988).
[10]Isaacs, S., and Holt, R. J., (Sept. 1987) Redefining procreation: Facing the issues, *Population Bulletin*, **42**, 30.
[11]Arditti, R. (Fall 1987) The surrogacy business, *Social Policy*, **18**, 44.
[12]*In the Matter of Baby M* 537 A.2d 1259, 1261 (N.J. 1988).
[13](September 6, 1986) Take the baby and run, *The Economist*, 27.
[14]*Surrogate Parenting v. Commonwealth ex. rel. Armstrong* 704 S.W. 2d 209 (Ky. 1986).

Bibliography

Andrews, L. (Oct./Nov. 1987) The aftermath of Baby M: Proposed state laws on surrogate motherhood, *Hastings Cent. Rep.* **17**(5), 31–40.

Annas, G. (June 1986) The baby broker boom, *Hastings Cent. Rep.* **16**(3), 30,31.

Arditti, R. (Fall 1987) The surrogacy business, *Social Policy* **18**, 42–46.

Bradley, T. (Summer 1987) Prohibiting payments to surrogate mothers: Love's labor lost and the constitutional right of privacy, *The John Marshall Law Review* **20**, 715–745.

Brahams, D. (Feb. 1987) The hasty British ban on commercial surrogacy, *The Hastings Cent. Rep.* **17**(1), 16–19.

Burt, R. (February 26, 1987) Judicial enforcement seen inappropriate, *New Jersey Law Journal* **119**, 24.

Capron, A. (Summer 1987) Alternative birth technologies: Legal challenges, *U.C. Davis Law Review* **20**, 679–704.

Fleming, A. (March 29, 1987) Our fascination with Baby M, *New York Times Magazine*, 33–37, 87.

Heyl, B. (March/April 1988) Commercial contracts and human connectedness, *Society* **25**, 11–16.

In RE Baby M 525 A.2d 1128 (N.J. Super. Ch. 1987).

In the Matter of BABY M 537 A.2d 1227 (N.J. 1988).

Isaacs, S., and Holt, R. J. (September 1987) Redefining procreation: Facing the issues, *Population Bulletin* **42**, 24–35.

Johnson, S. (Summer 1987) The Baby "M" decision: Specific performance of a contract for specially manufactured goods, *Southern Illinois University Law Journal* **11**, 1339–1348.

Morris, M. (March/April 1988) Reproductive technology and restraints, *Society* **25**, 16–21.

Neuhaus, R. (March/April 1988) Power, money and high-minded intentions, *Society* **25**, 28,29.

Neuhaus, R. (March/April 1988) Renting women, buying babies and class struggles, *Society* **25**, 8–10.

Roe v. Wade, 410 U.S. 113, 93 S. Ct. 705, 35 L.Ed. 2d 147 (1973)

Rothman, B. (March/April 1988) Cheap labor: Sex, class, race—and "surrogacy" *Society* **25**, 21–23.

Rothman B. (May/June 1987) Surrogacy: A question of values, *Conscience* **8**, 1–4.

Schwartz, L. (1987) Surrogate motherhood I: Responses to infertility, *The American Journal of Family Therapy* **15**, 158–162.

Stumpf, A. (November 1986) Redefining mother: A legal matrix for new reproductive technologies, *The Yale Law Journal* **96**, 187–208.

Surrogate Parenting v. Commonwealth ex. rel. Armstrong 704 S.W. 2d 209 (Ky. 1986)

Young, D. (1987) Surrogate motherhood legislation: A sensible starting point, *Indiana Law Review* **20**, 879–907.

Zelizer, V. (March/April 1988) From baby farms to Baby M, *Society* **25**, 23–28.

Surrogate Motherhood Agreements

The Risks to Innocent Human Life

James Bopp, Jr.

The New Jersey case, *In re Baby M.*,[1] forced upon the public consciousness the painful tensions of surrogate motherhood arrangements. The Office of Technology Assessment (OTA) of the United States Congress estimates in a recent report that 600 surrogacy arrangements have been made to date, in the United States alone.[2]

Various ethical and moral challenges have been mounted against surrogacy arrangements. However, the Office of Technology Assessment, in its report, *Infertility: Medical and Social Choices–Summary*, ignores such issues. Rather, the report, in keeping with the focus of the Office creating it, considers only the technological aspect, observing that "(s)urrogate motherhood is more a social solution to infertility than it is a medical technology."[3]

To deal with the social problems perceived in such arrangements, legislation has been introduced in over half of the state legislatures in the US.[4] Four states have passed statutes dealing with surrogacy.[5]

Since the goal of surrogacy arrangements is to produce children, there is a legitimate interest in protection of those children. The issue of risk to children produced by surrogacy arrangements may be formulated as follows:

Are there inherent risks to children, unborn or newly born, in surrogacy arrangements that cannot be overcome with appropriate legislation? It is the thesis of this article that the apparent potential risks to innocent human life can be resolved by appropriate legislation.[6]

Potential Risks to Innocent Human Life

Absent legislation, surrogacy agreements are governed by a contract among the natural father (who donates sperm for artificial insemination), the natural mother (who contributes her ovum and uterus), and her husband.[7] The relevant provisions of a typical contract, such as that employed in the *Baby M* case are as follows:

1. Forbid the natural mother to abort the fetus, unless her health is at risk or the unborn child is abnormal;
2. Require the surrogate mother to undergo amniocentesis or related prenatal tests for fetal normality; and
3. Provide for abortion upon the demand of the natural father if the unborn child is found to be defective.[8]

The potential risks to unborn and newborn children posed by such an agreement and by surrogacy arrangements in general are set forth below. Each will be dealt with to determine whether, given the present state of knowledge about surrogacy practices, the potential risks are surmountable with appropriate legislation, or are otherwise ameliorated by the nature of the agreement itself.

The Risk of Devaluation of Pregnancy, Childbirth, and the Unborn Child

One disturbing aspect of surrogacy arrangements is their treatment of the child as a chattel—a "thing," that can be bought, sold, and returned or even discarded, if found defective or undesirable.[9] Left unchecked, such an approach could foster a desensitization of society toward those who are physically or mentally handicapped. Moreover, surrogacy without controls could lead to a devaluation of the unborn child, its mother, and the process of pregnancy and childbirth.

One author described part of the danger as follows:

> Women willing to market their wombs have been seduced into honoring an androcentric vision, spawned by medical technologists, of pregnancy as a mechanical, extraneous event. By this vision, the fetus is an "insulated parasitic capsule," camped inside the gestator's pelvis—a pelvis that has been tidily sequestered from mind and spirit.[10]

The risk of desensitizing the public to the basic humanity of the unborn is analogous to the risk posed by those employing the Orwellian[11] euphemism of a "mass of cells" or "products of conception" for an unborn child, or denying that a "mother," "father," or "child" even exists until birth.[12] Unless this potential desensitization and devaluation can be dealt with by appropriate social controls through legislation recognizing and protecting the intrinsic value and dignity of all human life, including the unborn and the neonate, surrogacy would have to be opposed.

The Risk of Prenatal Tests for Fetal Perfection

Surrogacy contracts commonly require that the pregnant mother submit to amniocentesis or other prenatal tests. There are several problems with such tests.

One problem is that the tests may be extremely unreliable. The serum alphafetoprotein (AFP) test was shown in a California screening trial to be falsely positive in 95% of the cases and falsely negative in 22%.[13]

Because of the inherent problems and limited benefits of AFP testing, the American College of Obstetricians and Gynecologists (ACOG) advises that routine AFP screening "should not be implemented."[14] One writer notes that such tests are counterproductive in helping women make an informed choice:

> [U]nrestricted use...could increase the number of abortions of *normal* infants, minimize identification of affected infants, and heighten anxiety over the outcome of pregnancy [italics added].[15]

Amniocentesis sometimes indicates an abnormality that is proven erroneous upon examination following abortion.[16] One study shows amniocentesis has a 13% error rate in ascertaining fetal gender.[17] This would have a serious impact on reproductive choice and management when one natural parent is a known carrier of a sex-linked disorder.[15] Amniocentesis cannot distinguish between a fetus with a disability and one who is a mere carrier of genetic disease, and, therefore, abortion based on such test results would "deny life to a large number of unaffected offspring."[18]

Further, the Food and Drug Administration (FDA) has questioned whether sufficient amniocentesis services exist to bear a significant increase in the performance of the procedure.[19] Overburdening, as a result of large-scale use of surrogacy arrangements with prenatal testing requirements, could lead to further inaccuracies. One response could be an "entrepreneurial" burst of new services. This could result in lower quality levels, as one commentator has noted:

> [O]ne of the biggest technical problems is the receiving and communicating of accurate test results. Current laboratory services are overburdened and it has been predicted that a major increase in demand may

result in an unacceptable error rate. Also, because this is a potential $100 million industry, there is the additional concern that a profit may take precedence over quality control.[20]

As to the problem of counselor error, a commentator wrote:

> As is true for other kinds of diagnosis, the process of reproductive counseling is strewn with opportunities for missteps....A genetic counselor will be expected to exercise reasonable care to supply an accurate diagnosis....Nevertheless, a counselor is not required, any more than is any other medical practitioner, to supply error-free explanations.[21]

The most troubling aspect of fetal testing, such as the AFP screening program in California, is that such testing seems designed not to help the affected newborn, but to provide for the pregnancy's termination.[22] The notion of aborting children who are less than genetically or developmentally perfect is morally and ethically repugnant. Such unborn human lives have inherent worth apart from their perfection. In a wrongful life action, brought by the parents of a child with Down syndrome, who were not apprised of the availability of amniocentesis, the New Jersey Supreme Court, in *Berman v. Allan*, said:

> One of the most deeply held beliefs of our society is that life—whether experienced with or without a major physical handicap—is more precious than non-life....Concrete manifestations of this belief are not difficult to discover. The documents which set forth the principles upon which our society is founded are replete with references to the sanctity of life. The federal constitution characterizes life as one of three fundamental rights of which no man can be deprived of [sic] without due process of law....No man is perfect. Each of us suffers from some ailments or defects,

whether major or minor, which make impossible participation in all the activities the world has to offer. But our lives are not thereby rendered less precious than those of others whose defects are less pervasive or less severe.[23]

Moreover, children with handicaps are often able to lead productive and useful lives. Amniocentesis is unable to predict with certainty the degree of disability. Some relatively normal children, initially desired by parents, will be aborted because of the doubts and supposed severity of abnormalities suggested by screening.[24]

Finally, it should be noted that prenatal testing itself can be hazardous to unborn human life. A recent study of spontaneous abortions following chorionic villus sampling (CVS) found evidence that CVS results in a "small but significant risk of infection," which has been implicated in spontaneous abortions of chromasomally normal pregnancies. The overall loss rate was 5.4%, with infection found to be involved in 41% of these. Of those receiving CVS beyond 10 weeks following the last menstrual period, infection was involved in 69% of the losses.[25]

Amniocentesis also carries health risks. The risk of fetal or maternal injury is calculated at 1–2%, so that the risk of an abnormal fetus must be at least 2% before amniocentesis is "warranted" even on a eugenic basis.[26] However, the odds of a child being born with a disability, without amniocentesis being performed, are estimated at between one in 350 and one in 2000.[27] Risks to the fetus from amniocentesis include congenital abnormality, pregnancy complications, spontaneous abortion, neonatal death, and stillbirth.[28] Increased prenatal testing as a result of surrogacy agreement provisions is unreasonable and unjustified, would add to the risk to maternal health, and would sacrifice some unborn lives, including some that were not disabled. Unless mandatory prenatal screening in surrogacy arrangements can be eliminated legislatively, surrogacy arrangements involve an unacceptable risk of loss of unborn life.

The Risk of Paternally Mandated Abortion

Surrogacy contracts commonly call for the abortion of the unborn child upon the demand of the natural father in the event of genetic or congenital abnormalities in the fetus. The *Baby M* contract contained such a provision.[29] Such a provision may spring from the attitude on the part of the natural father and his wife that they are paying enough money to get a "perfect" child.[30]

Although such contract provisions are not likely to be found enforceable,[31] state legislatures should bar the inclusion of such provisions in surrogacy agreements, because of the potential coercive effect on women without legal counsel who might think they were legally required to abort because it said so in their surrogacy agreement.

The Risk of Maternally Elected Abortion

Another risk to innocent human life is that of maternally elected abortion. The typical contract requires that the surrogate mother not abort the child unless the child is disabled or there is risk to her health. Neither of these reasons is adequate to justify an abortion. Only if the risk to the mother reaches life-threatening proportions would the taking of the innocent, unborn human life be appropriate in order to save that of the mother.

A surrogate mother might decide she no longer wants to carry the child for a variety of reasons. For example, there might be a change of circumstances from a prior pregnancy to the present one. A surrogate is generally required to have already given birth to a healthy child of her own.[32] This is thought to decrease the possibility of her wanting to keep the child and also decreases somewhat the chance of the child born in her surrogate capacity being disabled. A surrogate may contract to bear a child with a certain burden of pregnancy in mind, similar to her previous one(s). However, in the event of a more difficult pregnancy, she might wish to abort the child.

Another problem might be the unanticipated stigma of being a surrogate mother. Social stigma has been a severe problem in some reported surrogacy arrangements.[33]

Although the contract may forbid the surrogate mother from obtaining an abortion for noncontractual reasons, it is questionable whether a woman may, by contract alone, be deprived of the right to choose an abortion, which was bestowed upon her by the United States Supreme Court in *Roe v. Wade*. Thus, a woman, whether pregnant for purposes of surrogacy or otherwise, is entitled to a virtually unlimited degree of abortion freedom under current constitutional interpretation.

For purposes of the present analysis, then, the relevant issue is whether surrogacy arrangements result in an increased number of unborn children at risk of abortion for less than life-threatening reasons. If this proves to be the case, and if this risk cannot be reduced by appropriate legislation, then surrogacy would have to be opposed as posing an unacceptable risk to the unborn.

The Risk of Nontreatment of a Disabled Newborn with Life-Threatening Ailments

A fourth readily apparent risk is that a child born in a surrogacy arrangement may be physically or mentally impaired and unwanted by either parent because of the perceived imperfection. If the child also has a life-threatening ailment, such as occurred in the Baby Doe case in Bloomington, Indiana,[34] there is a risk that the child will be neglected and not provided necessary life-saving medical treatment. If this risk proves substantial and legislatively insurmountable, then surrogacy arrangements are unacceptable because of this risk.

Analysis and Proposals Regarding Potential Risks

The following analysis presumes the continued viability of the Supreme Court's interpretation of the United States Constitution as set forth in *Roe v. Wade*. Were the Court to reconsider and alter its interpretation in *Roe*, or if a Human Life Amendment (protecting unborn human life) were passed, many of these concerns would be resolved. As the law now stands, certain constitutional limitations must be considered.

Constitutional Limitations

In *Roe v. Wade*,[35] the United States Supreme Court declared that a woman has a constitutional right, under the fourteenth amendment, to choose abortion free from state interference. Conceptually, this does not include protection from the asserted rights of others, such as the natural father of the unborn child. In two recent cases in Indiana, fathers of unborn children were able to obtain injunctions against the mother of their unborn child to prevent her from obtaining an abortion.[36] Each of these injunctions was granted after the trial court balanced the rights of the father in his child with the rights of the woman to have an abortion. Under such a theory, a court may enforce the father's rights in a surrogacy arrangement because his rights and interests, including the fact that the woman has contracted with him not to abort their child, outweigh her rights and interests in aborting the child.

It is important to note that the provisions of a surrogacy agreement are only binding insofar as the surrogate wishes to receive the promised fee and payment of her expenses for the pregnancy. If she chooses to abort the child, she loses the benefits of the contract and is in breach of contract. Money damages may be available to the natural father for her breach of contract, but he will not be able to force the woman to bear the child by obtaining judicial enforcement for specific performance of the contract, i.e., he cannot force her to carry the child to term on the sole basis of a contractual provision.

Thus, the bare existence of a contractual provision barring the woman from obtaining an abortion will not likely be an enforceable way to prevent such an abortion. The trial judge in *Baby M* ruled that the abortion right was not subject to contract.[37] The New Jersey Supreme Court, in reversing his decision, did not reach this issue.[38] However, some commentators feel that the right to choose abortion cannot be enforceably contracted away because of the fundamental nature of the right at issue.[39]

Of course, the easiest route to protection of the unborn in the surrogacy context would be for the state to bar surrogate mothers from obtaining abortions, except where the mother's life is at risk. This would be done under a bill pending in Minnesota.[40] Bills introduced in Hawaii[41] and Connecticut[42] would allow abortions only when necessary to protect the health of the mother, thus broadening the standard. All three proposals are more restrictive than current abortion jurisprudence, which virtually allows abortion on demand, for any or no reason, for the full nine months of pregnancy.[43]

Two commentators have suggested that such provisions, restricting a woman's right to abortion, probably violate *Roe*.[44] If this view is correct, a state could not presently bar women in surrogacy arrangements from obtaining an abortion.

Finally, although the resolution of complex constitutional matters, not yet settled in the courts, is beyond the scope of this article, it is possible that surrogacy arrangements will be found to be within the fundamental right to procreate or the right to privacy. If this were the case, then any state regulation of surrogacy arrangements would have to be narrowly tailored to meet a compelling state interest.[45] Because the states would be acting to protect weighty maternal, paternal, and fetal interests, it seems likely that the courts would find that the states had such compelling interests in regulation, and legislation carefully drafted to protect those interests would likely be upheld.

Analysis and Legislative Recommendations

Analysis of the potential risk factors and a recommended approach for legislation follows. This analysis is posited upon the assumption that, under current constitutional interpretation, restrictions on abortion will be considered unconstitutional.[46]

The Risk of Devaluation of Pregnancy, Childbirth, and the Unborn Child

The risk that the unborn child and the maternal process will be devalued seems capable of adequate amelioration under any of the legislative schemes currently proposed. The very act of passing comprehensive legislation to protect the interests at stake signals societal recognition of the value of those interests. Such legislation should have a purpose statement indicating that it is the intent to protect the unborn child to the limit of current constitutional interpretation.

The Risk of Prenatal Tests

Legislation could limit the risk of prenatal testing by making void any surrogacy agreements that require such testing. This would make it less likely that testing provisions would be inserted in such agreements. This would not entirely prevent the performance of prenatal testing, however. It might be possible to require constitutionally that no prenatal tests be done, but such a requirement would certainly be litigated and might be struck down as constitutionally unsound. A preferable route would be to make void contracts that include prenatal testing requirements and to require reporting of any prenatal testing and abortions performed in surrogacy settings. If there proves to be a statistically significant correlation between prenatal tests and abortions performed in surrogacy arrangements, it would then be necessary to make attempts to tighten the controls over these tests. It should be observed that women who are known carriers of genetic defects would not likely survive a basic screening for suitability to be a surrogate. There is no reason to believe that men with transmissible genetic defects would be more likely to employ a surrogate. Thus, there may actually be a somewhat reduced incidence of genetic abnormalities than in nonsurrogate parenting arrangements.

The Risk of Paternally Mandated Abortion

Any requirements for abortion demanded by the father should be prohibited from surrogacy agreements, and contracts containing them should be declared void by appropriate legislation. Under the Constitution, there is no legitimate way for a man to force a woman to abort her unborn child. No such interest or right exists on the father's part. However, under current surrogacy agreements, were she to refuse to abort on his demand, he could void the contract, and she would lose the consideration promised to her. This might create great pressure for her to abort the child, since she could still collect her promised fee (or a contractual portion thereof) by complying with his demand. Furthermore, women without legal counsel might believe that they had an obligation to abort because of the contract provision, not realizing that the provision could not be enforced. Thus, any surrogacy agreement requiring abortion on the father's mandate must be declared void, and such provisions must be banned before surrogacy arrangements may be considered acceptable.

The obvious solution in cases where the natural father does not want the child is to put it up for adoption upon birth. At least for the two most common congenital abnormalities, Down's syndrome and spina bifida, there is a waiting list of persons willing to adopt children with such disabilities.[47]

The Risk of Maternally Elected Abortion

Under current law, it is questionable, as has been noted, whether a woman may be legally prohibited from obtaining an abortion by a surrogacy contract. Legislatures, however, should enact such a provision until such provisions are determined definitively to be unconstitutional. State legislation should provide that, if a woman undergoes an abortion that is not necessary to save her life, the intended parents can recover any fees and expenses paid to the surrogate and reasonable attorney fees. Because such a provision is of questionable constitutionality, however, surrogate legislation should not be opposed without it.

Two state proposals provide such recovery if the surrogate undergoes an abortion that is not medically necessary.[48] Although this is not as strong as a life-of-the-mother provision, it is likely that few abortions will occur under even a "medically necessary" standard. This is so because of the inherent profitability of the surrogacy arrangement. Women who become involved in surrogacy generally do so out of a desire to be of assistance to a childless couple and to receive the monetary reward of such assistance. Neither of these motivations is likely to make surrogate women readily seek abortion. The social and psychological screening required by many state statutes is an important adjunct to the prevention of abortion and should be required in the surrogacy setting.

Whereas it is true that any abortions resulting from surrogate pregnancies would be a result of surrogacy arrangements, because the women would not have entered into these pregnancies apart from the arrangements, the risk of abortions in such settings seems sufficiently low that opposition to surrogacy should not be based on this risk alone. Nevertheless, reporting requirements should be enacted, and, if a significant number of abortions in surrogate pregnancies are discovered, then more stringent restrictions should be enacted. Then if more stringent restrictions were not constitutionally possible, surrogacy arrangements in general would have to be considered morally and ethically unacceptable. For the present, the risk does not seem significant.

The Risk of Nontreatment of a Defective Newborn with Life-Threatening Ailments

The risk that a physically or mentally impaired child with a life-threatening malady may be neglected or allowed to die from nontreatment may also be considered.

Surrogate Agreement Risks

The potential for such child abuse or infanticide is seen in the yet-unresolved case of *Malahoff v. Stiver*.[49] In *Malahoff*, a surrogate mother gave birth to a child with microcephaly in January 1983. She had contracted with Alexander Malahoff to be a surrogate mother for his child. Stiver felt no maternal bond, and Malahoff asked the hospital to withhold treatment from the child. In this case, the matter was resolved with the Stivers taking the child after Mr. Stiver was judicially found to be the legal father. No one had told the Stivers to abstain from sex during the time associated with the artificial insemination attempt.[50] Other cases might not be so easily and neatly resolved.

Any legislation permitting and regulating surrogacy must provide adequate protection for the unwanted newborn. Most of the legislative proposals set forth an explicit duty of the contracting couple to assume all parental responsibilities and rights upon the birth of the child, regardless of its condition. Such proposals are pending in California, the District of Columbia, Illinois, Maryland, Massachusetts, Michigan, Missouri, New Jersey, New York, Oregon, Pennsylvania, and South Carolina.[51] If such legislation is enacted and adequately enforced, then the risk to infants with a disability as well as a life-threatening condition should not be great enough to make surrogacy ethically and morally objectionable.

Conclusion

Given the present state of knowledge regarding surrogacy practice, it seems that the potential risks to innocent human life arising from surrogacy arrangements may be eliminated by appropriate legislation. However, reporting requirements should be enacted to ascertain the frequency of abortions or nontreatment decisions in surrogacy arrangements, and the results studied to be certain that unanticipated threats do not develop. For the present, there seems no reason why surrogacy arrangements governed by appropriate legislation should be deemed unacceptable based upon concern for the loss of innocent human life.

Notes and References

[1] 537 A.2d 1227 (N.J.1988).
[2] Office of Technology Assessment, *Infertility: Medical and Social Choices—Summary* 11 (1988).
[3] *Id.*
[4] *Id.*
[5] *Id.*
[6] There may be public policy reasons for opposing surrogacy arrangements. These are not dealt with here. The author addresses the issue of surrogacy only from the perspective of protecting innocent human life, both unborn and newly born.
[7] Brophy, K. M. (1981–82) Surrogate mother contract to bear children, *J. Fam. L.* **20,** 263, 266, 291. The natural mother is a "surrogate" only in the sense that she is a substitute for the natural father's wife, who, presumably, cannot bear children herself. Of course, there may be motivations for the natural father's wife to procure a surrogate other than nonfertility. The natural father's wife is not a party to the contract.
[8] In the *Baby M* case, Provision 13 of the contract declared:

> MARY BETH WHITEHEAD, Surrogate, agrees that she shall not abort the child....except, if in the professional medical opinion of the inseminating physician, such action is necessary for the physical health of MARY BETH WHITEHEAD or the child has been determined by said physician to be physiologically abnormal. MARY BETH WHITEHEAD further agrees, upon request of said physician, to undergo amniocentesis...or similar tests to detect genetic and congenital defects. In the event said test reveals that the fetus is genetically or congenitally abnormal, MARY BETH WHITEHEAD, Surrogate, agrees to abort the fetus upon demand of WILLIAM STERN, Natural

Father....If MARY BETH WHITEHEAD refuses to abort the fetus upon demand of WILLIAM STERN, his obligations as stated in this Agreement shall cease forthwith, except as to obligations of paternity imposed by statute. *In re Baby M*, 537 A.2d at 1268 (Appendix B).

[9]Johnson, S. H. (1987) The Baby "M" Decision: Specific performance of a contract for specially manufactured goods, *S. Ill. L.J.* **11**, 1339–1342, 1348 Bitner (1985) Womb for rent: A call for Pennsylvania legislation legalizing and regulating surrogate parenting agreements, *Dick. L. Rev.* **90**, 227, 235.

[10]O'Brien, S. (1987) The itinerant embryo and the neo-nativity scene: Bifurcating biological maternity, *Utah L. Rev.* **1**, 29, 33.

[11]In Orwell's *Oceania*, the Ministry of Truth (the Ministry of Propaganda) had three slogans: "War is Peace. Freedom is Slavery. Ignorance is Strength." G. Orwell, *1984* (1949). The official language, "Newspeak," likewise, attempted to restructure thought by redefinition and elimination of terms incompatible with the preferred orthodoxy. *Id.* (Appendix, "The Principles of Newspeak").

[12]It is noteworthy that, even in *Roe v. Wade*, 410 U.S. 113 (1973), the Supreme Court opinion that removed state-imposed barriers to women's freedom to choose abortion, employed both the term "unborn children" and "pregnant mother" in the course of its discussion of the competing interests. It also spoke of paternal interests, which it noted were not asserted, and, therefore, were not balanced by the Court against the pregnant mother's right to choose abortion. *Id.* at 165 n.67.

[13]Marmion, P. (1987) The California alpha-fetoprotein screening program, *Linacre Q* **54**, vol. 1 77, 87. Although followup tests are commonly used to reduce the risk of false negatives and positives, the risk remains substantial. Moreover, some women have been so frightened by the results of their first positive test "that they went off and obtained an abortion then and there." Marwick, C. (1983) Controversy surrounds use of tests for open spina bifida, *JAMA* **250**, 576, 577.

[14]ACOG Technical Bulletin #67, Oct., 1982. Annas, G. (Dec. 1985) Is a genetic screening test ready when the lawyers say it is? *Hastings Cent. Rep.* **15**, 6, 16.

[15]Nolan-Haley, J. M. (Spring 1982) Amniocentesis and human quality control, *Human Life Rev.* **8**, 51, 53; *see also* Marmion P., supra note 13.

[16]Friedman, J. M. (1974) Legal implications of amniocentesis, *U. Pa. L. Rev.* **123**, 92, 103; NICHD National registry for amniocente-

sis study group, (1976) midtrimester amniocentesis for prenatal diagnosis, *JAMA* **236,** 1471, 1472; Manganiello, P. D., Byrd, J. R., and McDonough, P. T. Tho. (1979) A report of the safety and accuracy of midtrimester amniocentesis at the Medical College of Georgia: Eight and one half years experience, *Am. J. Obstet. Gynecol.* **134,**911, 913. CF Editorial (1987) Routine prenatal genetic screening, *N. Engl. J. Med.* **317,** 1407, 1408 (noting that "serious errors" may be made in testing for Down syndrome with ultrasound screening).

[17]Freidman, J. M. supra note 16, at 103. A survey of the literature shows no reported improvement in sex determination.

[18]*Id.*

[19]Nolan-Haley, J. M., supra note 15, at 54.

[20]*Id.* at 55.

[21]Capron, A. M. (1979) Tort liability in genetic counseling, *Colum. L. Rev.* **79,** 618, 626–627.

[22]Marmion, P., supra note 13, at 82. *See also* references cited by Marmion.

[23]*Berman v. Allan*, 80 N.J. 421, 430, 404 A.2d 8, 12–13 (1979).

[24]Fletcher, J. C. (1978) Prenatal diagnosis: Ethical issues, in *Encyclopedia of Bioethics* (Reich, R., ed.) **3**, p. 1343.

[25]Wilson, Hogge, and Golbus (1987) Analysis of chromosomally normal spontaneous abortions after chorionic villus sampling, *J. Repro. Med.* **32,** 25.

[26]Friedman, J. M., supra note 16, at 105.

[27]*Hickman v. Group Health Plan, Inc.*, 396 N.W. 2d 10, 11, 14 (Minn. 1986); *see also*, Fineberg, K. S. and Peters, J. D. (Feb. 1984) Amniocentesis in medicine and law, *Trial* at 54, 59 (concerning major chromosomal abnormality, anencephaly, or spina bifida, "the odds that a 35- to 39-year-old woman will not have such a child are more than 200 to one without the test").

[28]Friedman, J. M., supra note 16, at 105.

[29]537 A.2d at 1268 (Appendix B).

[30]Note (1982) The surrogate motherhood contract in Indiana, *Ind. L. Rev.* **15,** 807, 822.

[31]*In re Baby M*, 217 N.J.Super. 313, 525 A, 2d 1128 (1987).

[32]Brophy, K. M., supra note 7, at 265.

[33]*Newsweek,* Jan. 19, 1987 at 49.

[34]*In re the Treatment and Care of Infant Doe*, No. GU-8204-004A, slip. op. (Monroe County Cir. Ct., Ind. April 12, 1983), Petition for Writ of Mandamus and Prohibition, denied sub nom. State of Indiana ex rel *Infant Doe v. Monroe*, 104 S. Ct. 394 (1983), reprinted in *Issues in Law and Medicine* **2,** 77 (1986). In this case, the child was born

with Down syndrome and blockage of the esophagus, which prohibited the intake of nutrients by mouth. Although the life-threatening blockage was easily remedied by routine surgery, the parents elected instead to allow the child to starve to death, presumably because they did not want a child with Down syndrome. Court appeals on behalf of the child were unavailing. Attempts to secure legal protection for neonates with disabilities by enforcement of §504 of the Rehabilitation Act of 1973, 29 U.S.C. §794, were struck down by a plurality of the U.S. Supreme Court in *Bowen v. American Hospital Association*. 106 S. Ct. 2101 (1986).

[35]410 U.S. 113.

[36]*Jane Doe v. John Smith*, No. 84 A01-8804-CV-00112, slip op. (Ind. App. 1 Dist. Oct. 24, 1988) (trial court reversed), *appeal denied*, No. 84A01-8804-CV112, slip op. (Ind. Feb. 9, 1989), *cert. denied*,— U.S.—(1989). The other was reversed by the Indiana Supreme Court and is now before the United States Supreme Court on appeal. *Conn v. Conn.* No. 73S01-8807-CV-631 (Ind. July 15, 1988) (Order affirming the court of appeals and adopting its opinion, reported at 525 N.E. 2d 612 (1988), *cert. denied*,—U.S.—(1989).

[37]217 N.J. Super. 313, 525 A.2d 1128.

[38]537 A.2d 1227.

[39]D'Aversa, C. Y. (1987) The right of abortion in surrogate motherhood agreements, *N. Ill. L. Rev.* **7**, 1; Note (1987) Rumplestiltskin revisited: The inalienable rights of surrogate mothers, *Harv. L. Rev.* **99**, 1936 (1986); Simons, K. W. (1980) Rescinding a waiver of a constitutional right, *Geo. L. J.* **68**, 919.

[40]H.F. 534 § 10(1) (i), Minn. (1983).

[41]H.B. 1009 § 9(a) (9), Haw. (1983).

[42]Comm. B. 5316 § 3(9), Conn. (1983).

[43]Although *Roe* posited limits the state might impose, in subsequent cases virtually all state regulations imposing even modest burdens on the woman seeking an abortion have been struck down. Moreover, the definition of "health" employed in *Doe v. Bolton*, 410 U.S. 179, 192 (1973), is so broad that a woman can readily find a physician to perform an abortion, even after viability, for a "health" reason.

[44]Note (1987) Surrogate motherhood legislation: A sensible starting point, *Ind. L. Rev.* **20**, 879, 895; Note (1986) Rumplestiltskin revisited: The inalienable rights of surrogate mothers, *Harv. L. Rev.* **99**, 1936.

[45]Note (1987) Surrogate motherhood legislation: A sensible starting point, supra note 44, at 880.

[46]A strong case may be made, although it is beyond the scope of

this paper, for the proposition that the presence of a surrogacy arrangement profoundly alters the nature and weight of the interests that the Supreme Court found compelling in *Roe* as a basis for bestowing constitutional protection to the abortion decision.

[47]The Spina Bifida Association of America, Rockville, MD, confirmed, in a phone conference on July 5, 1988, that there is a waiting list of persons wishing to adopt children born with spina bifida. The National Down Syndrome Congress, Park Ridge, IL, confirmed in a phone conference, on June 23, 1988, that there is currently a waiting list of persons desiring to adopt children with Down syndrome.

[48]*Id;* Note, Surrogate Motherhood, supra note 44, at 880.

[49]No. 83-6522 (E.D.Mich. 1988). Over 155 pleadings have been filed in support of claims and cross-claims, and both sides are awaiting the outcome of their respective motions for summary judgment. (Telephone conversation with judicial clerk, June 2, 1988).

[50]Andrews, L. B. (1984) The stork market: The law of the new reproductive technologies, *A.B.A.J.* **70,** 50, 56; *see also* D'Aversa, C. Y. (1987) The right of abortion in surrogate motherhood arrangements, *N. Ill. L. Rev.* **7,** 1, 31 n. 175.

[51]Andrews, L. B. (Oct./Nov. 1987) The Aftermath of Baby M, *Hastings Cent. Rep.* at 31, 37. California provides that, if the child has a disease or defect caused by the surrogate in violation of the contract, then the intended child-rearing couple need not take custody. *Id.*

The Vatican *Instruction* and Surrogate Motherhood

Richard Berquist

Introduction

The Vatican *Instruction on Respect for Human Life in Its Origin and on the Dignity of Procreation*,[1] in a brief section on public policy, puts a major emphasis on the role of politicians, but it also recognizes the need to develop an appropriate public consensus. It invites not only politicians, but also concerned men and women from many professions, to exercise a positive influence in this regard.

Consensus presupposes dialog. In this paper I will offer some reflections on the basis of ethical dialog, and then relate these reflections to the two main substantive issues raised by the *Instruction* in its section on public policy (Part III). These issues are (a) the right to life and physical integrity, and (b) the rights pertaining to marriage and the family, including the right of the child to be conceived, born, and raised by its parents. My intention is not to summarize the *Instruction*, but to suggest some lines of thought inspired by a study of the document and may help to facilitate dialogue on its public policy implications.

The Basis of Ethical Dialog

We may be tempted to despair about the possibilities of ethical dialogue, especially in so sensitive an area as artificial procreation. The incredibly fast pace of technological change and the complications of professional practice leave little time for reflection. How often one hears an overworked professional complain: "I've been so busy that I haven't had a chance to think about the ethical implications."

On the other hand, there is a widespread understanding of the need for ethical dialog, and some progress has been made. Looking to the future, I believe it would be helpful to distinguish three "levels" of ethical reflection. The first two levels are more obvious and have been extensively developed in the bioethics literature. The third level, though more difficult to understand, is of great importance for the study of controversial issues like those involved in artificial human procreation. I shall discuss these three levels briefly, using the problem of surrogacy as an illustration.

The first level of ethical analysis is the most familiar. It evaluates human actions by weighing and balancing what are perceived to be their desirable and undesirable consequences, seeking the best overall result. This level of analysis may be called consequentialism, and it includes utilitarianism in its various forms. Consequentialist arguments are often thought to be expressions of the "principle of beneficence."

In discussions of surrogate motherhood, first-level analysis is revealed by a sensitive concern for the physical, psychological, and financial burdens and benefits that may accompany or result from this practice. The analysis usually focuses primarily on the surrogate herself and on the other individuals closely involved in the transaction, but it may also take account of wider social burdens and benefits. The report of the Ethics Committee of the American Fertility Society (AFS) is noteworthy for the careful and detailed attention it has devoted to such matters.[2]

Though enormously useful, first-level analysis is not an adequate basis for ethical dialog. The inadequacy can easily be seen if we consider the difficulty of evaluating psychological suffering. Consider, for example, the period of grief that a surrogate sometimes experiences after giving up her baby. What is the significance of this suffering? Shall we treat it simply as a negative aspect of surrogacy, to be weighed against certain benefits; or is it an indication that something is fundamentally wrong with surrogacy itself? Could it be that surrogacy is not really consistent with the nature and dignity of motherhood? We can hardly ask such a question, let alone answer it, at this level of ethical reflection.

Furthermore, first-level analysis does not focus explicitly on the dignity of the human person. However, human dignity is surely the implicit basis of our ethical concern when we take account of the consequences of our actions on other people. Why would we be morally obligated to care about the welfare of other human beings if they had no more than utility value? The first level of analysis thus leads to a second level, which begins the process of understanding human dignity and its ethical implications more completely.

At the second level of ethical analysis, the consequences of our actions, though important, are not always decisive. The principle of respect for autonomy, i.e., the obligation to respect the right of persons to self-determination and free choice, must also be taken into account. Characteristic of this level of reflection is the extensive and often subtle literature on informed consent. The AFS Ethics Committee report gives careful attention to the factors that might compromise the free, informed consent of a potential surrogate mother. These include, for example, poverty, professional conflicts of interest, and pressure from friends and relatives.

Ethical reflection at this second level focuses more explicitly on human dignity than was possible or necessary at the first level. The principle of autonomy can be violated even when its violation causes no physical, psychological, or financial harm. Failure to obtain informed consent is morally wrong not only because it has undesirable consequences, but also because it violates the dignity of a rational, free being who exists for his or her own sake.

But even this second level of reflection is inadequate. Its most obvious limitation is its inability to distinguish between free choices that are reasonable and choices that are arbitrary. This becomes clear when we ask whether a person could ever freely choose to do or to undergo something personally degrading. Consider the question we raised earlier: Could a woman who chooses surrogacy violate, perhaps even without realizing it, the dignity of her own motherhood? At the second level of analysis, such a question seems almost a contradiction in terms.

Although the principle of free choice is rooted in human dignity, it reveals only one profile or dimension of that dignity. The human person is not merely a being with free choice; he or she has many other bodily and spiritual dimensions. All of these share in the dignity of the person. This brings us to the third level of analysis that considers what is suitable to the human person considered in his or her totality, with full respect for the objective significance, i.e., the intrinsic orientation and finality, of each of the various dimensions of human nature.

The third level of ethical analysis clarifies the ethical basis of the first two levels, places their arguments in perspective, and confirms or (when necessary) modifies or rejects their conclusions. It approves actions that produce, on balance, desirable consequences, but only if these actions are consistent with human dignity. Similarly, it respects freedom of choice (autonomy), but limits the exercise of this freedom according to ethical requirements rooted in a more complete understanding of human nature.

To clarify the meaning and importance of this third level of ethical analysis, I shall consider the right to life and the rights of marriage and the family as these pertain to artificial procreation. As noted earlier, these are the two issues that the Vatican *Instruction* singles out as particularly important for public policy. I hope that the discussion of these issues will also serve as a starting point for future dialog on the *Instruction* as a whole, because this document can be read with sophistication only if we are aware that its approach goes beyond consequentialism and autonomy to a full consideration of the implications of human dignity.

The Right to Life

The Vatican *Instruction* is concerned with the moral aspects of artificial procreation. It focuses on the question of whether artificial procreation is suited to the dignity of man and woman, and to the dignity of their child. It discusses the right to life because the practice of in vitro fertilization may be accompanied by the discarding or misuse of unwanted embryos.

The *Instruction* finds a subtle link between these two practices, IVF and the destruction of embryos. When procreation is removed from its natural context and placed under the control of technology, there is a tendency to think of human embryos as little more than the means to the achievement of pregnancy. With this goal in mind, some embryos are selected for transfer to the womb, and others are discarded or frozen (and thus subjected to risks not fully understood). Yet others may be chosen for experimentation to improve the success rate in future IVF procedures. Human beings become, in the words of the *Instruction*, the "giver of life and death by decree."

Although it is in some measure incidental to the main subject of the *Instruction*, the right to life would seem to be the document's most important concern from a public policy point of view, for this issue touches on the value of the human person most directly and fundamentally. If persons have dignity, i.e., if they exist for their own sake, then their lives are certainly worthy of respect. On the other hand, if human lives are *not* worthy of respect, what sense would it make to say that persons have dignity?

However, are embryos and preembryos persons? They do not seem human in the way that adults or even children and babies do. What seems insignificant tends to be treated as insignificant. Current American constitutional law, for example, does not recognize any right to life of an embryo or fetus at any time during the pregnancy. It is true that the state is permitted to prohibit abortions under certain circumstances after viability, but it is not required to do so. There are no constitutional rights belonging to the embryo or fetus *per se*.

Achieving consensus on the personhood of the embryo will not be easy. It seems clear enough that life begins at fertilization, but when does the person begin? By what criteria could the existence of the person be empirically verified or rejected? Some would argue, and I agree, that the embryo should be given the benefit of any doubt, but there are others who feel that there is no doubt that the embryo is definitely *not* a person.

Without going further into these questions, we can reasonably affirm that from the time of fertilization a new human person is at least emerging, if not actually present. (More than one person may in fact be emerging if twinning is destined to occur.) An emerging person is still undeveloped, yet not entirely so, since he or she is in the process of development. How are we to respect the emerging human person? This is a matter with which public policy must be concerned if it is to be sensitive to all the implications of human dignity.

There may be a closer connection than we have realized between the way we evaluate a fully developed person and the way we evaluate an emerging person. Each evaluation reflects on the other. To take an analogy: if Michaelangelo's *Pieta* is a great work of art, then the emergence of the *Pieta* was a process of great significance (even if no one realized this until the work was completed). By the same logic, if we were to deny the value of the emerging *Pieta*, we would have to deny the value of the *Pieta* itself. So it is with persons. The value of the person is reflected in the value of his or her emergence (and vice versa). The destruction of an embryo or fetus implies the insignificance of the person who was emerging in and through the process of embryonic and fetal development.

This line of argument is obviously not based on consequences (first-level reasoning) or on the principle of autonomy (second level). It proceeds explicitly from the objective intrinsic value of the person and belongs to the third level of ethical analysis.

The importance of the argument is confirmed by an analysis of the consequences of rejecting it. If the human person is insignificant when considered in his or her natural origin, then where does human dignity come from? It would seem to be a human invention—a creation of social evolution and human law. This in turn would cast doubt upon the foundations of democracy, which presupposes the equal personal dignity of all the citizens. If this equal dignity is no more than a human invention, on what basis, ultimately, could we respond to those who, like Thrasymachus in Plato's *Republic*, identify justice with the advantage of the stronger?

Our societal awareness of human dignity, already severely eroded by abortion, is further compromised by the practice of IVF. Just as abortion for sex selection violates the dignity of one of the sexes (adding to the indignity of abortion *per se*), and as the abortion of handicapped fetuses violates the dignity of handicapped persons, so the embryo destruction accompanying IVF violates human dignity in new ways. It is no longer just a matter of destroying embryos or fetuses who were never wanted; it is now a matter of deliberately creating them with the intention of destroying at least some of them, and perhaps even of using them in nontherapeutic experiments. In this way, a subhuman manner of evaluating embryonic life is woven ever more deeply into the fabric of social life.

The Rights of Marriage and the Family

The second issue emphasized by the Vatican *Instruction* in its section on public policy concerns "...the rights of the family and of marriage as an institution and, in this area, the child's right to be conceived, brought into the world and brought up by his/her parents."[3] These rights are founded on the morally necessary link between procreation and marriage, a link that is broken when donor gametes are used. Hence, the document specifically rejects donor insemination, embryo banks, and surrogate motherhood.

Procreation outside of marriage, like the destruction of embryos, is somewhat incidental to the main subject of the *Instruction*. Artificial procreation could, theoretically, be limited to married couples, using only their gametes, and procreation outside of marriage is usually natural, not artificial.

In fact, it is *natural* procreation outside of marriage that would seem to pose the more serious public policy question. It is far more frequent than artificial procreation, children are often unwanted, and the conditions for their proper upbringing are usually inadequate. In contrast, children conceived artificially by donor gametes are often desperately wanted and are carefully planned.

Nevertheless, artificial procreation with donor gametes may pose a greater threat in the long run to the family and the child's right to be conceived, brought into the world, and brought up by his or her parents. Because the child is wanted and provided for, artificial procreation can appear morally responsible. It directly challenges the idea of a necessary moral link between procreation and marriage. Natural procreation outside of marriage, on the other hand, usually appears irresponsible, since it often means that the couple has not formed a permanent relationship to carry out their joint responsibility for raising the child. Precisely because of its irresponsibility, natural procreation outside of marriage tends to reinforce our conviction that marriage is the appropriate context for having children.

Artificial procreation with donor gametes, which began as an attempt to cope with the problems of infertility within marriage, may in the end help to render marriage irrelevant. Once the moral link between procreation and marriage has been broken, extramarital procreation becomes, in principle, acceptable. The right to procreate then appears to belong to single persons and to homosexual and lesbian couples as well as to heterosexual couples. This implication may be approached cautiously, as in the AFS Ethics Committee report, or more boldly, as in the position of Barbara Katz Rothman.[4] However, it is difficult to avoid it altogether.

Whether we approve of it or not, artificial procreation outside of marriage is an important public policy issue. It bears within it the seeds of radical change. Of course, it seems almost inconceivable that the relationship of a mother and father to their natural children could ever be supplanted as the normal basis of family life. *Brave New World* is still fiction. Nevertheless, the prospect of a new social "ethos" in which heterosexual marriage would appear as just one among many child-raising options is full of uncertainties and potential dangers.

The link between procreation and marriage can be considered at each of the three levels of ethical reflection. Only the third level, however, would seem to provide a sufficiently fundamental context for appreciating all the dimensions of the problem. Let me try to illustrate this by showing how the question might be approached at each level.

The AFS Ethics Committee report provides a convenient example of first-level reasoning. It offers two consequentialist arguments in favor of linking child-rearing and marriage: (a) that it is helpful for children to have role models of both genders, and (b) that two parents can cope with the demands of child-rearing better than one parent alone. The Committee does not accept these arguments unreservedly, however, and it refuses to confine procreation to heterosexual marriage alone.[5]

A second-level approach would not emphasize the physical or psychological risks or benefits of surrogacy. It need not favor or oppose surrogacy *per se*. Instead, the focus would be on the principle of autonomy, understood as implying freedom of choice in reproductive matters, including the right to enter surrogacy arrangements. Within the narrow perspective of autonomy, an attempt to suppress surrogacy by insisting on a necessary moral link between procreation and marriage would seem to be a violation of human dignity.

If we approach the link between procreation and marriage at the third level of ethical reflection, new aspects of the question appear. I have argued that a third-level analysis must consider what is suitable to the dignity of the human person according to his or her nature, considered in all of its dimensions. Among these dimensions are the natural relationships of motherhood and fatherhood. Therefore, the obligation to respect human dignity fully includes an obligation to respect these natural relationships according to their essential meaning.

The origin of a new human being is from the germ cells of his or her natural parents. This is the foundation of the natural relationship of parents to their offspring, according to which a man and woman recognize a child as "theirs" in the most fundamental sense of the term. This relationship has implications. If human dignity requires that a new human being has the right to his or her full development as a person, then the natural parents have a joint responsibility for his or her upbringing. The initiation of any undertaking implies the responsibility to bring it to its appropriate fulfillment. When the undertaking concerns a new human being who exists for his or her own sake, the fulfillment of this responsibility is a clear moral obligation.

Corresponding to the parents' natural obligation to raise their child is the child's natural right to be raised by his or her parents. The child's right is a specific personal claim on the natural parents themselves; it is not directed to the community in general, at least not initially. Here we can find, I believe, the source of the moral link between procreation and marriage, if by marriage we understand a permanent relationship based on a solemn promise between a man and woman who cooperate in order to carry out their responsibility for the children who originate from them.

Sometimes, of course, the parents are unable or unwilling to carry out this responsibility. Then adoption may be necessary. Adoption does not deny the original responsibility of the natural parents. It is essentially a remedy, a way to provide for the welfare of the child when the child's right to be raised by his or her natural parents cannot be satisfied. This is radically different from the practice of sperm and ovum donation, in which the rights and responsibilities of natural parenthood are rejected in principle.

The importance of the child's natural right to be raised by his or her own parents can be seen even more clearly, if, for the sake of argument, we suppose the opposite, i.e., that the child does *not* have such a natural right. From this assumption, it would follow either that the child has no right to be brought up at all (which would certainly be opposed to human dignity) or that the child's right to be brought up correlates primarily with an obligation belonging to the community as a whole. However, if the community were primarily responsible for all children born, it would have the right to control reproduction. No one would have the right to procreate without government permission. Furthermore, the government would have to assign the task of raising children in accordance with what was perceived as the public interest. The natural parents would have the right to raise their own children only if this right were conceded them by the state.

Barbara Katz Rothman[6] argues for the right of the gestational mother to keep and to raise the child she has carried in her womb. Her argument bears a certain similarity to the one I have just developed, although it does not come to the same conclusion.

The similarity lies in the level of argumentation. Rothman does not attempt to found the right of the gestational mother on evidence that she would be the best qualified to raise the child or on any other sort of first-level (consequentialist) reasoning, nor does she rely on the principle of autonomy. She founds it on the gestational bond between mother and child, as though by the very fact of pregnancy a woman acquires a kind of natural right to the child—a right that takes precedence over legal arrangements like surrogacy.

Nevertheless, Rothman's conclusion seems mistaken. The gestational relationship, though necessary for the child's survival, is not the most fundamental link between mother and child. Gestation presupposes existence, and existence depends upon genetic motherhood and fatherhood. Therefore, the rights and responsibilities of parenthood must ultimately derive from the genetic link between parent and child.

There is some evidence, according to the AFS Ethics Committee report, that women are more willing to be surrogate carriers (providing only their wombs) than to be surrogate mothers (providing also their ova). This may indicate, the Committee cautiously concludes, that women feel they will be less attached to a child who is not genetically theirs.[7]

Conclusion

The Vatican *Instruction* is a controversial document. Its conclusions seem excessively "rigid" to many people and its reasoning difficult to grasp. Yet I believe it has something to teach us if we approach it in a spirit of dialog with an awareness of the different levels of ethical analysis that I have briefly described and illustrated above. If this essay can help to stimulate a more profound study of the *Instruction* and its public policy implications, it will have served its purpose.

References

[1] Congregation for the Doctrine of the Faith, *Instruction on Respect for Human Life and Its Origin and on the Dignity of Procreation* (March, 1987).

[2] American Fertility Society (Ethics Committee) (1986) Ethical considerations of the new reproductive technologies, *Fertil. Steril.* **46(3) Suppl. 1**, pp. 58–68.

[3] Congregation for the Doctrine of the Faith, *Instruction on Respect for Human Life and Its Origin and on the Dignity of Procreation* (March, 1987).

[4] Rothman, B. K., Recreating motherhood: Ideology and technology in contemporary society, this volume.

[5] American Fertility Society, p. 22.

[6] Rothman, B. K.

[7] American Fertility Society, p. 60.

Surrogacy

A Question of Values

Barbara Katz Rothman

People—friends, colleagues, maybe especially family—keep pointing out to me, with a bemused air, how strange it is to find me arguing on the same side as religious leaders in the debate on surrogacy arrangements. I too occasionally find some amusement in the "strange bedfellows" phenomenon. Indeed, with carefully groomed "happy surrogates" and their equally wellgroomed brokers placed on a TV or radio show by a professional public relations firm, I often find myself, side by side with some priest or rabbi, brought in by a producer to give a "balanced" view. So we argue the problems of surrogacy, sandwiched between car commercials, wine cooler ads, and other signs of the times.

The "tag" they use to identify me on television—the white line of print that shows up on the screen, but that I never get to see in the studio—sometimes reads "author," rarely reads "sociologist," but most often reads "feminist." So there we are, "feminist" and "priest" or "rabbi" arguing the same, antisurrogacy side.

*Reprinted with permission from *Conscience* **7(3)**, 1–4 (May/June 1987).

However, it is only on the very surface that I am on the same side as these religious leaders. We may have landed on the same side of this particular fence, but we have taken very different paths to get here, and we are headed in very different directions. The values that I use in my opposition to surrogacy are fundamentally different from those the church is using, and the goals I seek are just as different. Strangely enough, in many ways my values are much the same as those used by those of my feminist colleagues who have come to an opposite conclusion. So rather than just presenting conclusions—the deeply damaging nature of so-called "surrogacy" arrangements as I see it—I want to use this opportunity to explore these issues of values.

The arguments against surrogacy that come out of traditional religious contexts most often rest on two basic principles: first, that surrogacy is "unnatural" because it goes against the nature of women and especially of mothers; and second, that it violates the sanctity of the family. Feminists are a lot less sure about just what is "natural" for women, but on the whole, we have concluded that the institution of motherhood as it exists in our society is pretty far from any natural state. Feminists are not about to get caught up in any "maternal instincts" arguments. Women end pregnancies with abortions, or end their motherhood by giving a baby up for adoption, when that is what they feel they need to do. Being pregnant does not necessarily mean a woman is going to mother and raise the child that might be born of that pregnancy, nor are we going to claim that only a birth mother can mother and nurture a baby. We know that loving people, men as well as women, can provide all the warm, caring, loving nurturance a baby needs.

Feminists are also not concerned with maintaining the "sanctity of the family," a pleasant-enough-sounding phrase that has been used to cover an awful lot of damage. That was the argument offered to allow men to beat their wives and children, the argument used to stop funding day care centers, and the argument used most generally to stop women from controlling their own lives and their own bodies. The "family" whose "sanctity" is being maintained is the patriarchal, male-dominated family. Feminists have a different sense of family; we need to protect the single young mother and her child, the lesbian couple and their children, and the gay man's family. As feminists, we are concerned not with the control, ownership, and kinship issues of the traditional family, but with the *relationships* that people establish with one another, both with adults and with children.

Why then, as feminist, do I oppose the surrogacy relationship? The "liberal" wing of feminism does not necessarily oppose these contracts. As long as the women entering into them do so of their own volition, with fully informed consent, and as long as they maintain their control over their own bodies throughout the pregnancies, some feminists have said that surrogacy contracts should be supported by the state. Some go as far as Lori Andrews, for instance, a noted liberal feminist, who says that these contracts should be binding, with absolutely no opportunity for the mother to change her mind. Andrews says that it is important for women to be held to *these* contracts if women are ever to be taken seriously in legal contracts.

What then are the objections that I, as a more radical feminist, as a socialist feminist, raise to surrogacy? More importantly, on what values—values more basic to me than "a deal's a deal"—am I basing my objections?

My values place relationships as central. Rather than the ownership or kinship ties that appear to epitomize the "traditional" or patriarchal family, I value the interpersonal relations people establish. A man does not own his wife, nor does he have ownership rights over the child she produces of her body. Men may own their sperm, but children are not sperm grown up. Children are not "owned," and they are certainly not to be available for sale. On the other hand, children do not enter the world from Mars or out of a black box. Children, as it says in the books for children, come from mothers. They enter the world in a relationship—a physical, social, and emotional relationship with the woman in whose body they have been nurtured. The nurturance of pregnancy is a relationship, one that develops as a fetus becomes more and more a baby.

That does not mean that the maternal relationship cannot be ended, nor does it mean that the relationship is the most overwhelming, all-powerful relationship on earth. In fact, we know it to be a fairly fragile relationship. The intimacy that a mother and her baby experience can be easily lost if they are separated. If a woman chooses to end this relationship, so be it. When a mother chooses to give a baby up to others who want to raise that baby as their own, she is doing what we have all done in our lives: ending a relationship. Sometimes this is done with less, and sometimes with more pain, but rarely is it an easy thing for a mother to do. The relationship that a woman has established by the time she births her baby has more weight, in my value system, than claims of genetic ties, of contracts signed, or of down payments made.

When I make this argument with traditionally oriented people, I am often asked if this doesn't contradict my ideas about a woman's right to abort an unwanted pregnancy. Not at all, I think. When a woman chooses an abortion, she is choosing not to enter into a maternal relationship. Women want access to safe abortions as quickly as possible, before a relationship can be begun. In sum, in my value system, I am placing the woman, her experiences and her relationships, at the very heart of my understanding of all pregnancies.

The second value I bring as a feminist to my understanding of surrogacy contracts is the value of women's bodily autonomy, control over their bodies—and I see the fetus as part of a woman's body. Traditional patriarchal values would see the fetus as part of its father's body—his "seed" planted in a woman's body. In a patriarchal system, the father—or the state or church—are held to have special control over a woman's body and life because of the fetus that she can bear. As a feminist, I reject that. The fetus is *hers*. Women never bear anybody else's baby: not their husbands', not the state's, and not the purchaser's in a surrogacy contract. Every woman bears her *own* baby. I believe that to be true regardless of the source of the sperm, and regardless also of the source of the egg.

Most of the surrogacy arrangements we have seen so far, like the Stern–Whitehead case, were done not with elaborate new reproductive technology, but with the very old and very simple technology of artificial insemination. The "surrogate" has been the mother in every possible way to the baby she bore, but some newer technologies allow eggs to be transferred from woman to woman—allowing a woman to be pregnant with a fetus grown of another woman's egg.

The Church rejects technology. I do not. When technology is used to allow a woman to enter motherhood with a pregnancy, in just the way that artificial insemination has been used to allow an infertile man to become a social father beginning with a pregnancy, I have no problem with it. My concern is when that technology is used in a so-called surrogacy arrangement—when the birth mother (the pregnant woman) is declared to be only a "rented womb" or a "surrogate," and the "real" mother is declared to be the woman who donated the egg. By making it possible for black and brown women to be used to bear white babies, this technology will bring the costs of surrogacy down and the controls on surrogates up. Some of us fear the development of "baby farming", with babies being produced on a "Holly Farms" or "Perdue" model.

The Church objects to the technology because the Church values the fertilized egg itself as an object it believes to be human. I object when women are "used," when parts of women are put up for sale or hire, and when our relationships are discounted in favor of genetic ties and monetary ties.

The liberal feminists who would allow surrogacy contracts still demand rights of bodily autonomy for the so-called surrogate. The compromise position that they maintain is that it is indeed her body, and she must have all decision-making control over her pregnancy, but it is not her baby in her body if she has contracted it away. I feel that that kind of "compromise" does a profound disservice to women. I cannot ever believe that a woman is pregnant with someone else's baby. That idea is repugnant to me—it reduces the woman to a container. Also, I do not think that that kind of compromise—saying the pregnancy is indeed hers, but the fetus/baby theirs (the purchasers)—can be workable. The "preciousness" of the very wanted, very expensive baby will far outweigh the value given to the "cheap labor" of the surrogate.

We are encouraging the development of "production standards" in pregnancy—standards that will begin with the hired pregnancy, but grow to include all pregnancies. This is the inevitable result of thinking of pregnancy not as a relationship between a woman and her fetus, but as a service she provides for others, and of thinking of the woman herself not as a person, but as the container for another, often more valued, person.

So, yes, I agree with the patriarchal religious leaders who say that the state should not recognize surrogacy arrangements, but no, we're not really on the same side. Strange world, isn't it?

Surrogacy and the Family

*Social and Value Considerations**

Adrienne Asch

The topic of surrogate parenting arouses strong feelings. The Baby M case and its aftermath stirred powerful emotions in millions made aware of the drama through newspaper, magazine, or television coverage of the trial, or, through the made-for-TV movie. Unexpectedly faced with the need to come up with law and policy on surrogacy, state legislatures may act hastily, without examining how surrogacy and other reproductive arrangements challenge us to articulate shared values about the meaning of family life, parenthood, sexuality, and reproduction.

Formulating policy on surrogacy requires that lawmakers and the general public attempt to ask and answer many questions that have hitherto gone unasked. Theory and research in the social sciences provide some guidance in answering them. Ultimately, however, the major questions for our society concern moral values and policy directions, and not questions answerable through social science theory or research. The reflections that follow raise more questions than they answer. They identify issues

---

*An earlier version of this article was presented as a "Staff Discussion Draft" to the Task Force on New Reproductive Practices of the New Jersey Commission on Legal and Ethical Problems in the Delivery of Health Care at its meeting of March 16, 1988. The views expressed herein reflect those of the author and not those of the Commission or its Task Force on New Reproductive Practices.

that must be addressed in the policy discussions that must take place if we are to contemplate surrogacy thoughtfully.

Value of the Family

Friendship and love, marrying, and having children exemplify some of the tangible and intangible goods of life thought to be outside of and not subject to purchase or sale.[1] The reasons stem from their special character. Ideally, the decisions to marry and procreate spring from people's deepest feelings and beliefs. Friendship and love develop from the sharing and self-disclosure that grow out of and reinforce bonds of respect and affection.

The acts of marrying and having children, and the children themselves are special, in part because they symbolize and embody people's deepest longings for closeness and commitment to others. Despite divorce, turbulent parent–child relationships, and apparent increases in domestic violence in the nation's families, and despite criticism from all parts of the political and cultural spectrum about the current state of family values and government family policy, scholars in the field conclude that virtually all adults consider family the most important source of meaning and value in their lives. Berger and Berger[3] report that 92% of adults rank it first, ahead of friendship, work, patriotism, and religion. Even if families no longer fulfill economic functions, and sometimes require outside assistance to care for some of their members, society still looks to families to supply what Berger and Berger describe as sources of meaning, value, and identity. Families are expected to provide nurturance, security, affection, and stability for adults and children.

Surrogacy compels us to examine what we consider the essence of family life, what values we seek to uphold, and how much deviation we can tolerate from what has been a cultural ideal. To be a "family," must an adult heterosexual married couple raise children? Do women, as some feminists contend,[4] find themselves resorting to new technologies to have children in response to pressure from men or society rather than in response to their own desires? Is marriage or a long-term adult couple relationship incomplete without children?

If family is viewed as the best place for raising children, and if children can receive the optimum stability and love in a family, what type of family should it be? Surrogacy offers two groups of people the opportunity to have children: the group most often discussed (and by far the largest) consists of the heterosexual couple rendered involuntarily childless by medically definable infertility;[†] the other group consists of those for whom childlessness results from social circumstance—adults without partners or adults with homosexual partners who seek to add children to their lives.

Until the advent of assisted reproduction, infertile couples who could not adopt remained without children—a situation often extraordinarily painful to them and touching deep sympathies of the approximately 85% of couples who are parents.[5] With the advent of technology enabling reproduction without heterosexual sex, we face the possibility of increasing numbers of single adults and homosexual couples seeking medical assistance and/or state recognition to realize their desires to fulfill what psychologists consider a normal developmental task of giving something to a new generation by being parents.[6] These practices cause us to consider under what conditions and in what forms of family we believe children should be raised.

[†]This may also include diagnosis of secondary infertility, arising when a couple has already produced children, but cannot have another by ordinary means.

Although Berger and Berger are strong advocates of what they describe as "bourgeois family" for socializing children into independent, freedom-loving adults, they comment that the nation will tolerate ethnic, religious, and class diversity in families, if those families inculcate prized virtues of honesty, industry, and self-reliance in their young. These authors make no comment about whether they believe the nation will or should tolerate a multiplicity of single-parent households or homosexual households, although they offer no evidence to suggest that such households are in themselves inadequate for children.

What we may, however, be willing to tolerate if done privately is one thing, and what we may wish to endorse by any type of state action and assistance is another. Our task is to articulate what we prize about adult intimacy that might lead to parenthood, and to spell out the values and type of family constellations that will be best for children.

Along with exploring that set of policy questions, it will be important to consider the meaning and possible legal and policy implications of what has been termed a "right to procreate" and the psychological meaning of parenthood. Exactly what is meant when proponents of surrogacy point out that the practice enables women and men to fulfill their "right to procreate" or right to found a family? If such rights exist, are they synonymous? What category of social problem is addressed when people speak of a problem in founding a family? Does the longing to raise a child entail a corresponding right to have one to raise if biological or social circumstances preclude doing so?

When legal scholars and court opinions have endorsed a right to procreate, traditionally they have referred to the right of an individual to make reproductive

Surrogacy and the Family

decisions without interference from others—whether the "others" were husbands seeking to block their wives from having abortions; doctors performing sterilizations on welfare mothers without their knowledge or consent; or administrators of state institutions ordering sterilizations of residents classified as retarded or mentally ill. The right to procreate was understood as a right not to be interfered with. As Stumpf [7] notes, it has been defined more as a right not to procreate or as a right to preserve one's procreative potential.

Like other proponents of surrogacy as a reproductive alternative, Stumpf argues that the right to procreate, understood as a right to be free of interference or coercion, leads to a positive right to reproduce. Proponents of surrogacy—as a means of fulfilling the positive right to procreate—have drawn two inferences from the existing negative right. First, they argue that the right of a person not to be interfered with in reproductive decisions entails a positive right of a person to reproduce; second, they claim a person's right to obtain assistance from another person, perhaps even from society's laws and institutions, to make reproductive arrangements that will involve someone other than themselves. They would turn a negative right —the right of a person not to be interfered with—into a positive right—the right of a person to obtain assistance from other parties and from the laws of the state to fulfill their desire to have children.

These inferences are worth questioning. Since baby-making still requires the genetic contribution of a male and a female, neither a man's nor a woman's desire to reproduce entails a corresponding societal obligation to assist his or her efforts to find a satisfactory reproductive method or partner if biological or social circumstances have failed to afford such an opportunity.

Stumpf does not persuade me that preserving an individual's right to procreate necessarily entails derivative rights and obligations to reshape laws and social structures to assist people in doing so. She and others do, however, give voice to the psychological and social importance of parenthood in adult life, and to the longings of infertile people to participate in the experiences of causing a child to come into the world and of raising that child. Neither Erikson[8] nor other researchers in adult psychological development would claim that parenthood is the only way that an adult can fulfill desires to be generative and to give something to the next generation. Nonetheless, many adults believe that their relationships with their children are the most fulfilling part of their lives.[8] Hoffman and Hoffman[9,10] and Gerson[11] report that adults expect that having children will provide opportunities to give and receive love, affirm their status as adults in their families and society, meet obligations stemming from their religion or identification with an ethnic group, and aid their potential for psychological growth. The desire for such manifold enrichment does not cease with the discovery of infertility or with the recognition that one's social situation precludes ordinary reproduction.

The new reproductive practices provide the opportunity, and perhaps the necessity, for examining what is viewed as essential about parenthood—psychologically, socially, morally, and legally. How do we and should we accord value to the genetic, gestational, and social aspects of parenthood? How should we understand the psychological and social meaning of infertility, the desire to have a genetically related child, or the commitment to a nonbiological child on the part of a waiting would-be parent?

Because the overwhelming majority of adults who raise children are both their biological and social parents, we unfortunately have little data on such topics as the comparative weight that today's adults attach to the genetic, gestational, and social aspects of parenting. Whether states choose to countenance or to discourage surrogacy arrangements, they will have to decide how to resolve disputes of custody and parental rights if the surrogate mother changes her mind about relinquishing the child. In addition to assessing the parent's ability to meet the child's needs using a "best interest" standard or a fitness standard, the state will have to consider whether it should use any adult-centered standard as part of its method of resolving custody disputes. If so, the law will have to measure a father's genetic contribution and intent to raise and nurture a child against a mother's genetic contribution and experience of carrying and bearing a child. Furthermore, as Stumpf [12] points out, the intent to raise a child manifested on the part of an infertile wife who participates with her husband in undertaking surrogacy deserves some social, moral, and perhaps legal recognition. As discussed below, such moral and legal evaluations must be made with incomplete information about the psychological and social significance of the pregnancy period on prospective mothers or fathers.

Consequences for Children and Adults

Whatever else it should do, family policy should aim to protect the most vulnerable people in the society—generally, children. How will the practice of surrogacy affect the resulting children, the siblings or half-siblings of those children, and parent–child relationships in the families of surrogate mother and biological father?

Potential harm to adults must also be considered. If the ideal of couples as parents is to share in the conception as well as the raising of children, what is the effect on the marital relationship and the parent–child relationship when one parent has a genetic tie to a shared child and one does not? Does the nonbiological parent suffer in the marital and family relationship? Do spouses of surrogates suffer? If we learn that they do, what weight should be accorded their stress? What is the psychological impact on the surrogate of selling or giving reproductive material or services to create a life for which she will not be responsible and with which she will not be involved after birth?

If we believe that the psychological consequences to the participants in surrogacy should matter in formulating appropriate public policy, we must ask two questions: First, by what standard should we judge psychological harm? Given that children and adults already adapt to a host of family arrangements, what is the ideal we espouse and against which we measure families? Second, how much weight should we give to evidence of potential psychological harm in formulating law and policy? These questions require empirical work as well as value clarification about the importance of psychological well-being for the good of individuals and society. Concerns about potential psychological effects must be weighted against values of autonomy and preserving what we consider fundamental to our understanding of family, sexuality, and reproduction.

The Lessons and Limitations of Analogies

Some supporters and opponents of surrogacy draw on psychological theory and research to buttress arguments for or against the practice. There are difficulties, however, in using some of the existing theory and data to draw conclusions about surrogacy. Commentators cite data available on pregnancy, parent–child interaction after birth, impact of adoption on children, impact of relinquishment

Surrogacy and the Family

on birth parents, and participants in donor insemination to illustrate the pros and cons of surrogacy.

Research on the hormonal changes of pregnancy has been used to argue that the birth mother is undoubtedly more connected to and committed to the child at its birth than any other adult and should be presumed to have custody of the child in case of dispute between her and the father. The research on pregnancy has been conducted on women who intended to keep their children after birth. There is no research comparing the physical and psychological experiences of pregnancy of women who keep their babies to the experiences of those who relinquish their babies to adoption or to surrogacy arrangements, or to those of women who plan to do so and change their minds. Thus, data on pregnancy as a physical and psychological preparation for parenthood have been obtained only on women who intended to rear the children they bore.

Surrogacy's supporters and opponents cite findings about the impact of adoption on children. Discovering that more adoptees than nonadoptees receive inpatient or outpatient mental health treatment, and finding that adoptees are more vulnerable to learning difficulties, behavioral disorders, and other symptoms of psychological distress, some opponents of surrogacy argue that children are harmed by growing up in families in which they are raised by one or more unrelated adult. Supporters of surrogacy counter that most adoptees are well within normal ranges on most measures of psychological adjustment, and that most of the difficulties are confined to middle childhood and adolescence. Growing up in nontraditional families thus causes no adverse consequences for the vast majority of children, and none that persist into adulthood.

Whatever the preponderance of data shows about the impact of adoption on children, the applicability of such findings to the phenomenon of surrogacy is open to question. There is no reason to think that the genetic endowment, prenatal history, or postnatal environment of the typical adoptee and the typical child of surrogacy will be comparable. Nor is there any way to predict whether a child of surrogacy will experience social stigma, trauma on discovering the truth about their origins, or anxiety and uncertainty about identity that comes from ignorance about their biological parents, as have many adoptees. The child of surrogacy may be stigmatized, depending on the climate of public opinion. By contrast, if surrogacy is practiced with openness and information is available about the birth mother (as is generally agreed to by the parties prior to conception), the child of surrogacy will not have to manage without information, or face the anxiety and uncertainty of a search for roots, as is typical for the adoptee whose quest for biological identity is blocked by parents lacking information and by sealed agency records.

Literature on adoption is only now sorting out the proportion of adoptee problems attributable to the secrecy, stigma, problematic prenatal histories, and other sequelae of adoption. If surrogacy continues, it may be practiced in ways thought to avoid the problems of children growing up in adoptive and other nontraditional families. However, depending upon one's values and perspective, one can weigh the evidence of problems of adoptees in either of two ways: as severe enough to discourage creating any more children than are absolutely necessary with discontinuities in their biological and social parentage, or as showing that the difficulties of adoptees are only somewhat greater than those of nonadoptees on some measures at certain life states, and thus that the harm is not so severe as to warrant a ban on bringing children into the world who will have at least one adoptive parent.

Studies of the postadoption experiences of relinquishing mothers have been relied upon to support the argument that depriving surrogate mothers of the children they bear is likely to lead to severe and prolonged emotional difficulty. Such reports as that of Deykin, Campbell, and Patti[13] are cited and interpreted to show that relinquishing women experience grief, anxiety, and unresolved feelings of guilt and depression for up to 30 years after they surrender their children. Relinquishment data may fit the situation of the surrogate who is unwilling to give up the child, but may very poorly fit the situation of the typical surrogate. Usually, the surrogate is older than the woman who surrenders her child to adoption; is already raising a child of her own; has entered into the surrogacy agreement aware from the outset that she does not wish to raise the child she is carrying; expects from the outset that she will relinquish the child; and furthermore, is assured of knowing something about the people who will care for the child, and is generally assured that the child will have access to information about her if the child wishes it.

Given the findings of the relinquishment data that birth mothers were plagued for years by lack of information about their children's welfare, and given some data reporting that obtaining information about their children relieved much anxiety,[14] it may be that surrogacy policy and practice can learn from the mistakes made in adoption. If surrogacy is to continue, some of the problems that secrecy causes for relinquishing parent, adoptee, and adoptive parent can be avoided.

The discussion of the uses of theory and data in making policy on surrogacy suggests, that although cautious use of data may be worthwhile, the existing data are problematic and their applicability to surrogacy uncertain.

Thus, clarity about social values will be of as much aid to the policymaker as awareness of facts.

Consequences for Society

In addition to evaluating the possible consequences of surrogacy upon the parties directly involved, it is most important to determine possible consequences for fundamental social values: promoting gender equality and wider options for both sexes; treating the body with dignity; harming our understanding of intimacy and sexuality; and changing our thinking about the value of children.

From Shulamith Firestone to Margaret Atwood, we have a host of different perceptions about which reproductive arrangements will maximize and which will impede equality for women and men, both as parents and as participants in the world beyond home and children. Feminists committed to fostering equality for women and men in and out of the home disagree on whether surrogacy promotes or undermines gender equality.

Feminist opposition to surrogacy and other reproductive techniques has been more vocal than feminist support. Although responses differ based upon the particular practice under discussion, prevalent views emphasize how the techniques pressure infertile women into using medical professionals to have children they might otherwise not have; how surrogacy exploits poor women who participate for lack of other means to support themselves and their children; how the techniques have been developed by male scientists to enable men to have children with their genes, regardless of women's wishes; how technology medicalizes pregnancy and increasingly puts it under professional (equated with male) control; how fostering genetic and particularly gestational surrogacy will

Surrogacy and the Family

increase the tendency to link women with reproduction and thwart efforts to offer women and men ways of seeing women outside of their capacity to bear children; and ultimately, how new reproductive practices will enable men to have children through extrauterine reproduction, depriving women of their role in creating offspring.[15]

Other feminist voices have argued that new reproductive arrangements can empower both women and men to decide how, whether, and under what conditions they will be biological or social parents.[16] These authors present thoughtful and powerful arguments to show circumstances in which surrogacy can be positive and affirming for women, and they argue that their decisions to give up children could be understood to be genuinely informed decisions.

Prevalent in Western religion and in many cultures have been views of the deep personal significance of the human body, its parts and acts, and of sexuality, reproduction, creating life, and the children created. It may not be clear whether we attach significance to children and creating life because they are the results of intimate physical and sexual acts, or whether at least some of the significance we attach to our bodies and sexuality is caused by its link to reproduction—to new lives. Creating public policy toward surrogacy requires articulating shared cultural understandings of the links we wish to maintain between having children and sexual and personal relationships.

As Murray[17] discusses, many religions and cultures have construed the human body and its parts as being of personal significance, and as inseparable from the dignity and worth to be accorded the human soul or humans as persons. Because reproduction involves genetic and bodily material and gestational services, as a matter of religious and cultural thought and traditional social policy, it could continue to be deemed an act that should take place ideally for reasons intrinsic to the dignity and spirit of the people involved in the creation of a child. Opponents of commercial surrogacy plausibly might contend that one adult cannot purchase life-giving genetic material or gestational services from another without viewing the seller as a means to an end, thus turning a personal genetic and gestational contribution to a shared enterprise of creating life into a market transaction. By this argument, commercial surrogacy (or perhaps even sale of eggs and sperm) is viewed as harmful to respect for persons or to the appreciation of people as unique individuals.

This argument—focusing on the payment for the contribution and service of the adult involved—does not apply to criticisms of noncommercial surrogacy, whether between strangers or friends and relatives. If creating new life is profound because it requires using our bodies in intimate sexual acts, reproducing when the two people involved have no prior or ongoing personal and sexual relationship, and no commitment to share in the raising of the new life, may diminish the meaning attached to the act of creating life and to the new child created. Those criticisms could stem from a conviction that, ideally, a child should be created only by two people who are committed to contributing to the child's life, not merely biologically, but as psychological and social parents as well. Donating blood or even a kidney to preserve existing life might be seen, both morally and in policy terms, as different from selling or even giving genetic material or reproductive services to create life for which one intends to take no ongoing responsibility. The former might be seen as fostering social solidarity by increasing people's sense of responsibility toward one another; the latter could be viewed as weakening a sense of responsibility for and involvement with others.

Surrogacy and the Family

Even though assisted reproduction—such as commercial or noncommercial surrogacy—is reproduction without sex, deciding what policy to adopt toward it may require that we articulate what shared beliefs we hold about the desirable connection of sexuality and reproduction, and possibly about the place of sexuality in adult life. A majority of the nation believes that sexuality as an expression of intimacy is desirable and moral, even when birth control and abortion are used to prevent sexual expression by adults from leading to unintended children. Nonetheless, surrogacy is reproduction without sexuality, by two people who may have no relationship with each other for any other purpose than obtaining money for one and a child for the other. Public policy may not wish to encourage creating children in a way that so departs from understandings of the connection of children with sexual and personal relationships.

Even if the adult who allows services or material to be purchased or donated believes his or her own ends and life plan are enhanced by using his or her body in such a way, the effect on the children created and on cultural understandings of the value of children in adult lives must also be examined. Obviously, many children learn that they resulted from accidents, from chance encounters, or from their parents' desire for sexual intimacy, and not necessarily for a child. Does the child's recognition of the circumstances of its birth in itself constitute a harm to that child? Again, recognizing that existing children already at some time grasp the reality that their birth may not have been desired, should the state condone a system that permits children to be created so that one of the biological participants creates for pay or to satisfy a sense of altruism?

Common objections to baby-selling arise both because baby-selling exploits women's hardships and because it puts price tags on the child—the new human life. Children have been valued both as unique and separate beings, and as tangible embodiments of the adults who created them. Surrogacy might or might not harm our common perception of the value of children to adults as unique and separate people—precious regardless of the circumstances of their creation, simply because they are human beings. It might, however, alter (for good or ill) our appreciation of children as ideally stemming from and enhancing the relationships and lives of the two adults who were their biological progenitors.

Children deserve to be valued for many reasons, and social policy should protect them and foster appreciation of them. Whether children created through surrogacy will experience harms to their own sense of self-worth, and whether our social understanding of the worth of children will diminish, remain questions primarily of value, and only secondarily of fact.

References

[1]Walzer, M. (1983) *Spheres of Justice* (Basic Books, New York).

[2]Radin, M. J. (June, 1987) Market-inalienability, *Harvard Law Review* 100, 1839–1937; Walzer, M., *Ibid.*

[3]Berger, B. and Berger, P. L. (1983) *The War over the Family: Capturing the Middle Ground* (Anchor/Doubleday, New York); Skolnik, A. S. (1983) *The Intimate Environment*, Third Ed., Little Brown, Boston.

[4]Arditti, R., Klein, R. D., and Minden, S. (eds.) (1984) *Test-Tube Women: What Future for Motherhood?* (Pandora Press, Boston).

[5]Office of Technology Assessment (1988) *Infertility: Medical and Social Choices;* Skolnick, A. S. (1983) *The Intimate Environment*.

[6]Erikson, E. H. (1969) *Identity and the Life Cycle* (W. W. Norton, New York); Erikson, E. H. (1982) *The Life Cycle Completed* (W. W. Norton, New York).

[7]Stumpf, A. E. (1986) Redefining motherhood: A legal matrix for the reproductive technologies, *Yale Law Journal* 96, 187–208.

[8]Erikson, E. H. (1969); Erikson, E. H. (1982).

[9]Hoffman, L. and Hoffman, M. (1973) The value of children to parents, in *Psychological Perspectives in Population* (Basic Books, New York), pp. 19–76.

[10]Hoffman, *Ibid.*

[11]Gerson, M. J. (1983) A scale of motivation for parenthood: The index of parenthood motivation, *J. Psychol.* 113, 211–220; Gerson, M. J. (1986) The prospect of parenthood for women and men, *Psychology of Women Quarterly* 10(1), 49–62.

[12]Stumpf, A. E.

[13]Deykin, E., Campbell, L., and Patti, P. (1984) The post-adoption experience of surrendering parents, *Am. J. Orthopsychiatry* 54, 271–280.

[14]Silverman, P., Campbell, L., Patti, P., and Style, C. (1988) Reunions between adoptees and birthparents: The birthparents' experience, *Social Work* 33(6), 523–528.

[15]Arditti, R., Klein, R. D., and Minden, S. (eds.); Chavkin, W., Rothman, B. K., and Rapp, R. (1989) Alternative modes of reproduction: Other views and questions, *Reproductive Laws for the 1990s* (Humana Press, Clifton, New Jersey), pp. 405–409; Spallone, P. and Steinberg, D. L. (eds.) (1987) *Made to Order: The Myth of Reproductive and Genetic Progress* (Pergamon, New York).

[16]Andrews, L. B. (1989) Alternative modes of reproduction, *Reproductive Laws for the 1990s* (Humana Press, Clifton, New Jersey), pp. 361–403; Purdy, L. (1989) Surrogate mothering: Exploitation or empowerment, *Bioethics* 3(1), 18–34; Zipper, J. and Sevenhuijsen, S. (1987) Surrogacy: Feminist notions of motherhood reconsidered, in *Reproductive Technologies: Gender, Motherhood, and Medicine* (University of Minnesota Press, Minneapolis), pp. 118–138.

[17]Murray, T. H. (1987) On the human body as property: The meaning of embodiment, markets, and the needs of strangers, *University of Michigan Journal of Law Reform* 20(4), 2055–2088.

Appendices

Appendices

SURROGATE PARENTING AGREEMENT

THIS AGREEMENT is made this 6th day of February, 19 85, by and between MARY BETH WHITEHEAD, a married woman (herein referred to as "Surrogate"), RICHARD WHITEHEAD, her husband (herein referred to a "Husband"), and WILLIAM STERN, (herein referred to as "Natural Father").

RECITALS

THIS AGREEMENT is made with reference to the following facts:

(1) WILLIAM STERN, Natural Father, is an individual over the age of eighteen (18) years who is desirous of entering into this Agreement.

(2) The sole purpose of this Agreement is to enable WILLIAM STERN and his infertile wife to have a child which is biologically related to WILLIAM STERN.

(3) MARY BETH WHITEHEAD, Surrogate, and RICHARD WHITEHEAD, her husband, are over the age of eighteen (18) years and desirous of entering into this Agreement in consideration of the following:

NOW THEREFORE, in consideration of the mutual promises contained herein and the intentions of being legally bound hereby, the parties agree as follows:

1. MARY BETH WHITEHEAD, Surrogate, represents that she is capable of conceiving children. MARY BETH WHITEHEAD understands and agrees that in the best interest of the child, she will not form or attempt to form a parent-child relationship with any child or children she may conceive, carry to term and give birth to, pursuant to the provisions of this Agreement, and shall freely surrender custody to WILLIAM STERN, Natural Father, immediately upon birth of the child; and terminate all parental rights to said child pursuant to this Agreement.

2. MARY BETH WHITEHEAD, Surrogate, and RICHARD WHITEHEAD, her husband, have been married since 12/2/73, and RICHARD WHITEHEAD is in agreement with the purposes, intents and provisions of this Agreement and acknowledges that his wife, MARY BETH WHITEHEAD, Surrogate, shall be artificially inseminated pursuant to the provisions of this Agreement. RICHARD WHITEHEAD agrees that in the best interest of the child, he will not form or attempt to form a parent-child relationship with any child or children MARY BETH WHITEHEAD, Surrogate, may conceive by artificial insemination as described herein, and agrees to freely and readily surrender immediate custody of the child to WILLIAM STERN, Natural Father; and terminate his parental rights; RICHARD WHITEHEAD further acknowledges he will do all acts necessary to rebut the presumption of paternity of any offspring conceived and born pursuant to aforementioned agreement as provided by law, including blood testing and/or HLA testing.

3. WILLIAM STERN, Natural Father, does hereby enter into this written contractual Agreement with MARY BETH WHITEHEAD, Surrogate, where MARY BETH WHITEHEAD shall be artificially inseminated with the semen of WILLIAM STERN by a physician. MARY BETH WHITEHEAD, Surrogate, upon becoming pregnant, acknowledges that she will carry said embryo/fetus(s) until delivery. MARY BETH WHITEHEAD, Surrogate, and RICHARD WHITEHEAD, her husband, agree that they will cooperate with any background investigation into the

Surrogate's medical, family and personal history and warrants the information to be accurate to the best of their knowledge. MARY BETH WHITEHEAD, Surrogate, and RICHARD WHITEHEAD, her husband, agree to surrender custody of the child to WILLIAM STERN, Natural Father, immediately upon birth, acknowledging that it is the intent of this Agreement in the best interests of the child to do so; as well as institute and cooperate in proceedings to terminate their respective parental rights to said child, and sign any and all necessary affidavits, documents, and the like, in order to further the intent and purposes of this Agreement. It is understood by MARY BETH WHITEHEAD, and RICHARD WHITEHEAD, that the child to be conceived is being done so for the sole purpose of giving said child to WILLIAM STERN, its natural and biological father. MARY BETH WHITEHEAD and RICHARD WHITEHEAD agree to sign all necessary affidavits prior to and after the birth of the child and voluntarily participate in any paternity proceedings necessary to have WILLIAM STERN'S name entered on said child's birth certificate as the natural or biological father.

4. That the consideration for this Agreement, which is compensation for services and expenses, and in no way is to be construed as a fee for termination of parental rights or a payment in exchange for a consent to surrender the child for adoption, in addition to other provisions contained herein, shall be as follows:

(A) $10,000 shall be paid to MARY BETH WHITEHEAD, Surrogate, upon surrender of custody to WILLIAM STERN, the natural and biological father of the child born pursuant to the provisions of this Agreement for surrogate services and expenses in carrying out her obligations under this Agreement;

(B) The consideration to be paid to MARY BETH WHITEHEAD, Surrogate, shall be deposited with the Infertility Center of New York (hereinafter ICNY), the representative of WILLIAM STERN, at the time of the signing of this Agreement, and held in escrow until completion of the duties and obligations of MARY BETH WHITEHEAD, Surrogate, (see Exhibit "A" for a copy of the Escrow Agreement), as herein described.

(C) WILLIAM STERN, Natural Father, shall pay the expenses incurred by MARY BETH WHITEHEAD, Surrogate, pursuant to her pregnancy, more specifically defined as follows:

(1) All medical, hospitalization, and pharmaceutical, laboratory and therapy expenses incurred as a result of MARY BETH WHITEHEAD'S pregnancy, not covered or allowed by her present health and major medical insurance, including all extraordinary medical expenses and all reasonable expenses for treatment of any emotional or mental conditions or problems related to said pregnancy, but in no case shall any such expenses be paid or reimbursed after a period of six (6) months have elapsed since the date of the termination of the pregnancy, and this Agreement specifically excludes any expenses for lost wages or other non-itemized incidentals (see Exhibit "B") related to said pregnancy.

(2) WILLIAM STERN, Natural Father, shall not be responsible for any latent medical expenses occurring six (6) weeks subsequent to the birth of the child, unless the medical problem or abnormality incident thereto was known and treated by a physician prior to the expiration of said six (6) week period and in written notice of the same sent to ICNY, as representative of WILLIAM STERN by certified mail, return receipt requested, advising of this treatment.

(3) WILLIAM STERN, Natural Father, shall be responsible for the total costs of all paternity testing. Such paternity testing may, at the option of WILLIAM STERN, Natural Father, be required prior to release of the surrogate fee from escrow. In the event WILLIAM STERN, Natural Father, is conclusively determined not to be the biological father of the child as a result of an HLA test, this Agreement will be deemed breached and MARY BETH

WHITEHEAD, Surrogate, shall not be entitled to any fee. WILLIAM STERN, Natural Father, shall be entitled to reimbursement of all medical and related expenses from MARY BETH WHITEHEAD, Surrogate, and RICHARD WHITEHEAD, her husband.

(4) MARY BETH WHITEHEAD'S reasonable travel expenses incurred at the request of WILLIAM STERN, pursuant to this Agreement.

5. MARY BETH WHITEHEAD, Surrogate, and RICHARD WHITEHEAD, her husband, understand and agree to assume all risks, including the risk of death, which are incidental to conception, pregnancy, childbirth, including but not limited to, postpartum complications. A copy of said possible risks and/or complications is attached hereto and made a part hereof (see Exhibit "C").

6. MARY BETH WHITEHEAD, Surrogate, and RICHARD WHITEHEAD, her husband, hereby agree to undergo psychiatric evaluation by JOAN EINWOHNER, a psychiatrist as designated by WILLIAM STERN or an agent thereof. WILLIAM STERN shall pay for the cost of said psychiatric evaluation. MARY BETH WHITEHEAD and RICHARD WHITEHEAD shall sign, prior to their evaluations, a medical release permitting dissemination of the report prepared as a result of said psychiatric evaluations to ICNY or WILLIAM STERN and his wife.

7. MARY BETH WHITEHEAD, Surrogate, and RICHARD WHITEHEAD, her husband, hereby agree that it is the exclusive and sole right of WILLIAM STERN, Natural Father, to name said child.

8. "Child" as referred to in this Agreement shall include all children born simultaneously pursuant to the inseminations contemplated herein.

9. In the event of the death of WILLIAM STERN, prior or subsequent to the birth of said child, it is hereby understood and agreed by MARY BETH WHITEHEAD, Surrogate, and RICHARD WHITEHEAD, her husband, that the child will be placed in the custody of WILLIAM STERN'S wife.

10. In the event that the child is miscarried prior to the fifth (5th) month of pregnancy, no compensation, as enumerated in paragraph 4(A), shall be paid to MARY BETH WHITEHEAD, Surrogate. However, the expenses enumerated in paragraph 4(C) shall be paid or reimbursed to MARY BETH WHITEHEAD, Surrogate. In the event the child is miscarried, dies or is stillborn subsequent to the fourth (4th) month of pregnancy and said child does not survive, the Surrogate shall receive $1,000.00 in lieu of the compensation enumerated in paragraph 4(A). In the event of a miscarriage or stillbirth as described above, this Agreement shall terminate and neither MARY BETH WHITEHEAD, Surrogate, nor WILLIAM STERN, Natural Father, shall be under any further obligation under this Agreement.

11. MARY BETH WHITEHEAD, Surrogate, and WILLIAM STERN, Natural Father, shall have undergone complete physical and genetic evaluation, under the direction and supervision of a licensed physician, to determine whether the physical health and well-being of each is satisfactory. Said physical examination shall include testing for venereal diseases, specifically including but not limited to, syphilis, herpes and gonorrhea. Said venereal disease testing shall be done prior to, but not limited to, each series of inseminations.

12. In the event that pregnancy has not occurred within a reasonable time, in the opinion of WILLIAM STERN, Natural Father, this Agreement shall terminate by written notice to MARY BETH WHITEHEAD, Surrogate, at the residence provided to the ICNY by the Surrogate, from ICNY, as representative of WILLIAM STERN, Natural Father.

13. MARY BETH WHITEHEAD, Surrogate, agrees that she will not abort the child once conceived except, if in the professional medical opinion of the inseminating physician, such action is necessary for the physical health of MARY BETH WHITEHEAD or the child has been determined by said physician to be physiologically abnormal. MARY BETH WHITEHEAD further agrees, upon the request of said physician to undergo amniocentesis (see Exhibit "D") or similar tests to detect genetic and congenital defects. In the event said test reveals that the fetus is genetically or congenitally abnormal, MARY BETH WHITEHEAD, Surrogate, agrees to abort the fetus upon demand of WILLIAM STERN, Natural Father, in which event, the fee paid to the Surrogate will be in accordance to Paragraph 10. If MARY BETH WHITEHEAD refuses to abort the fetus upon demand of WILLIAM STERN, his obligations as stated in this Agreement shall cease forthwith, except as to obligations of paternity imposed by statute.

14. Despite the provisions of Paragraph 13, WILLIAM STERN, Natural Father, recognizes that some genetic and congenital abnormalities may not be detected by amniocentesis or other tests, and therefore, if proven to be the biological father of the child, assumes the legal responsibility for any child who may possess genetic or congenital abnormalities. (See Exhibits "E" and "F").

15. MARY BETH WHITEHEAD, Surrogate, further agrees to adhere to all medical instructions given to her by the inseminating physician as well as her independent obstetrician. MARY BETH WHITEHEAD also agrees not to smoke cigarettes, drink alcoholic beverages, use illegal drugs, or take non-prescription medications or prescribed medications without written consent from her physician. MARY BETH WHITEHEAD agrees to follow a prenatal medical examination schedule to consist of no fewer visits than: one visit per month during the first seven (7) months of pregnancy, two visits (each to occur at two-week intervals) during the eighth and ninth month of pregnancy.

16. MARY BETH WHITEHEAD, Surrogate, agrees to cause RICHARD WHITEHEAD, her husband, to execute a refusal of consent form as annexed hereto as Exhibit "G".

17. Each party acknowledges that he or she fully understands this Agreement and its legal effect, and that they are signing the same freely and voluntarily and that neither party has any reason to believe that the other(s) did not freely and voluntarily execute said Agreement.

18. In the event any of the provisions of this Agreement are deemed to be invalid or unenforceable, the same shall be deemed severable from the remainder of this Agreement and shall not cause the invalidity or unenforceability of the remainder of this Agreement. If such provision shall be deemed invalid due to its scope or breadth, then said provision shall be deemed valid to the extent of the scope or breadth permitted by law.

Appendices

19. The original of this Agreement, upon execution, shall be retained by the Infertility Center of New York, with photocopies being distributed to MARY BETH WHITEHEAD, Surrogate and WILLIAM STERN, Natural Father, having the same legal effect as the original.

_____   2/6/85
WILLIAM STERN   DATE
Natural Father

STATE OF New York)
) SS.:
COUNTY OF New York)

On the 6 day of February, 19 85, before me personally came WILLIAM STERN, known to me, and to me known, to be the individual described in the foregoing instrument and he acknowledged to me that he executed the same as his free and voluntary act.

_____
NOTARY PUBLIC

DONNA SPISELMAN
Notary Public, State of New York
No. 41-4792053
Qualified in Queens County
Commission Expires March 30, 19 85

We have read the foregoing five pages of this Agreement, and it is our collective intention by affixing our signatures below, to enter into a binding legal obligation.

Mary Beth Whitehead   1-30-85
MARY BETH WHITEHEAD, Surrogate   DATE

Richard Whitehead   1-30-85
RICHARD WHITEHEAD   DATE
Surrogate's Husband

STATE OF New York)
) SS.:
COUNTY OF New York)

On the 6th day of February, 1985, before me personally came MARY BETH WHITEHEAD, known to me, and to me known to be the individual described in the foregoing instrument and she acknowledged to me that she executed the same as her free and voluntary act.

Donna Spiselman
NOTARY PUBLIC
DONNA SPISELMAN
Notary Public, State of New York
No. 41-4732053
Qualified in Queens County
Commission Expires March 30, 1985

STATE OF New York)
) SS.:
COUNTY OF New York)

On the 6 day of February, 19 85, before me personally came RICHARD WHITEHEAD, known to me, and to me known to be the individual described in the foregoing instrument and he acknowledged to me that he executed the same as his free and voluntary act.

Donna Spiselman
NOTARY PUBLIC
DONNA SPISELMAN
Notary Public, State of New York
No. 41-4732053
Qualified in Queens County
Commission Expires March 30, 1985

Appendices

Act No. 199

Public Acts of 1988
Approved by the Govemor
June 27, 1988
Filed with the Secretary of State
June 27, 1988

STATE OF MICHIGAN
84TH LEGISLATURE
REGULAR SESSION OF 1988

Introduced by Senators Binsfeld, Welborn, Gast, Cropsey, Carl, DiNello, Ehlers, Dillingham, Mack, Barcia, J. Hart, Vaughn, Cruce, Dingell, and DeGrow

ENROLLED SENATE BILL No. 228

AN ACT to establish surrogate parentage contracts as contrary to public policy and void; to prohibit surrogate parentage contracts for compensation; to provide for children conceived, gestated, and born pursuant to a surrogate parentage contract; and to provide for penalties and remedies.

The People of the State of Michigan enact:

Sec. 1. This act shall be known and may be cited as the "surrogate parenting act".

Sec. 3. As used in this act:
(a) "Compensation" means a payment of money, objects, services, or anything else having monetary value except payment of expenses incurred as a result of the pregnancy and the actual medical expenses of a surrogate mother or surrogate carrier.

(b) "Developmental disability" means that term as defined in the mental health code, Act No. 258 of the Public Acts of 1974, being sections 330.1001 to 330.2106 of the Michigan Compiled Laws.

(c) "Mental illness" means that term as defined in the mental health code, Act No. 258 of the Public Acts of 1974.

(d) "Mentally retarded" means that term as defined in the mental health code, Act No. 258 of the Public Acts of 1974.

(e) "Participating party" means a biological mother, biological father, surrogate carrier, or the spouse of a biological mother, biological father, or surrogate carrier, if any.

(f) "Surrogate carrier" means the female in whom an embryo is implanted in a surrogate gestation procedure.

(g) "Surrogate gestation" means the implantation in a female of an embryo not genetically related to that female and subsequent gestation of a child by that female.

(h) "Surrogate mother" means a female who is naturally or artificially inseminated and who subsequently gestates a child conceived through the insemination pursuant to a surrogate parentage contract.

(i) "Surrogate parentage contract" means a contract, agreement, arrangement in which a female agrees to conceive a child through natural or artificial insemination, or in which a female agrees to surrogate gestation, and to voluntarily relinquish her parental rights to the child.

Sec. 5. A surrogate parentage contract is void and unenforceable as contrary to public policy.

Sec. 7. (1) A person shall not enter into, induce, arrange, procure, or otherwise assist in the formation of a surrogate parentage contract under which an unemancipated minor female or a female diagnosed as being mentally retarded or as having a mental illness or developmental disability is the surrogate mother or surrogate carrier.

(2) A person other than an unemancipated minor female or a female diagnosed as being mentally retarded or as having a mental illness or developmental disability who enters into, induces, arranges, procures, or otherwise assists in the formation of a contract described in subsection (1) is guilty of a felony punishable by a fine of not more than $50,000.00 or imprisonment for not more than 5 years, or both.

Sec. 9. (1) A person shall not enter into, induce, arrange, procure, or otherwise assist in the formation of a surrogate parentage contract for compensation.

(2) A participating party other than an unemancipated minor female or a female diagnosed as being mentally retarded or as having a mental illness or developmental disability who knowingly enters into a surrogate parentage contract for compensation is guilty of a misdemeanor punishable by a fine of not more than $10,000.00 or imprisonment for not more than 1 year, or both.

Appendices

(3) A person other than a participating party who induces, arranges, procures, or otherwise assists in the formation of a surrogate parentage contract for compensation is guilty of a felony punishable by a fine of not more than $50,000.00 or imprisonment for not more than 5 years, or both.

Sec. 11. If a child is born to a surrogate mother or surrogate carrier pursuant to a surrogate parentage contract, and there is a dispute between the parties concerning custody of the child, the party having physical custody of the child may retain physical custody of the child until the circuit court orders otherwise. The circuit court shall award legal custody of the child based on a determination of the best interests of the child. As used in this section,"best interests of the child" means that term as defined in section 3 of the child custody act of 1970, Act No. 91 of the Public Acts of 1970, being section 722.23 of the Michigan Compiled Laws.

Sec. 13. This act shall take effect September 1, 1988.

This act is ordered to take immediate effect.

....
Secretary of the Senate.

....
Clerk of the House of Representatives.

Approved...

....
Governor.

Excerpts from **Nevada Revised Statutes,**
Chapter 773, Sec. 6(5) (1987)

Sec. 6.1. Except as otherwise provided in subsection 3, it is unlawful for any person to pay or offer to pay money or anything of value to the natural parents of a child in return for the natural parent's placement of the child for adoption or consent to or cooperation in the adoption of the child.

2. It is unlawful for any person to receive payment for medical and other necessary expenses related to the birth of a child from a prospective adoptive parent with the intent of not consenting to or completing the adoption of the child.

3. A person may pay the medical and other necessary living expenses related to the birth of a child of another as an act of charity so long as the payment is not contingent upon the natural parent's placement of the child for adoption or consent to, or cooperation, in the adoption of the child.

4. This section does not prohibit a natural parent from refusing to place a child for adoption after its birth.

5. The provisions of this section do not apply if a woman enters into a lawful contract to act as a surrogate, be inseminated and give birth to the child of a man who is not her husband.

The New York State Task Force on Life and the Law Statement

Recommendations look at surrogacy in medical, legal, social context

The following summarization is a verbatim reprint of the conclusions and recommendations of the New York State Task Force on Life and the Law regarding surrogate parenting:

Part 1: The Medical, Legal, and Social Context

- Surrogate parenting is not a technology, but a social arrangement that uses reproductive technology (usually artificial insemination) to enable one woman to produce a child for a man and, if he is married, for his wife. Surrogate parenting is characterized by the intention to separate the genetic and gestational aspects of child-bearing through an agreement to transfer the infant and all maternal rights at birth

- The well-publicized "Baby M" case has given surrogate parenting a prominent place on the public agenda. Nonetheless, the reproductive technologies used in the arrangements—artificial insemination and, increasingly, in vitro fertilization—also pose profound questions about the ethical, social, and biological basis of parenthood. In addition, the procedures to screen donors raise important public health concerns. The task force will address those issues in its ongoing deliberations and recognizes that they form part of the context within which surrogate parenting must be considered.

- Legal questions about surrogate parenting, although novel in many respects, arise within the framework of a well developed body of New York family law. In particular, policies about surrogate parenting will necessarily focus on two basic concerns in all matters involving the care and custody of children—the protection of the fundamental right of a parent to rear his or her child and in the promotion of the child's best interests.

- The Supreme Court of New Jersey has ruled that paying a surrogate violates state laws against baby selling. Surrogacy agreements may also be found invalid because they conflict with comprehensive statutory schemes that govern private adoption and the termination of parental rights.

- In New York, it is uncertain whether surrogate parenting contracts are barred by the statute that prohibits payments for adoption. If not, it is probable that the surrogate could transfer the child to the intended parents by following private adoption procedures. If a dispute about parental rights arises before the surrogate consents to the child's adoption, custody would probably be determined based on the child's best interests. Regardless of the outcome, the court ordinarily will have no basis for terminating the parental status of either the surrogate or the intended father.

• The right to enter into and enforce surrogate parenting arrangements is not protected as part of the constitutional right to privacy. Surrogate parenting involves social and contractual—rather than individual—decisions and arrangements that may place the rights and interests of several individuals in direct conflict. The commercial aspects of surrogate parenting also distinguish the practice from other constitutionally protected private acts. Constitutional protection for the right to privacy is diminished when the conduct involved assumes a commercial character.

• The social and moral issues posed by surrogate parenting touch upon five central concerns: (1) individual access and social responsibility in the face of new reproductive possibilities, (2) the interests of children, (3) the impact of the practice on family life and relationships, (4) attitudes about reproduction and women, and (5) application of the informed consent doctrine.

• Surrogate parenting has been the subject of extensive scrutiny by public and private groups, including governmental bodies in the US and abroad, religious communities, professional organizations, women's rights organizations, and groups that advocate on behalf of children and infertile couples. Of the governmental commissions that have studied the issue, many concluded that surrogate parenting is unacceptable. In this country, six states have enacted laws on surrogate parenting, four of which declared surrogate contracts void and unenforceable as against public policy.

Part II: Deliberations and Recommendations of the Task Force

• As evidenced by the large body of statutory law on custody and adoption, society has a basic interest in protecting interests of children and in shielding gestation and reproduction from the flow of commerce.

• When surrogate parenting involves the payment of fees

and a contractual obligation to relinquish the child at birth, it places children at risk and is not in their best interests. The practice also has the potential to undermine the dignity of women, children, and human reproduction.

- Surrogate parerenting alters deep-rooted social and moral assumptions about the relationship between parents and their children. The practice involves unprecedented rules and standards for terminating parental obligations and rights, including the right to a relationship with one's own child. The assumption that "a deal is a deal", which is relied upon to justify this drastic change in public policy, fails to respect the significance of the relationships and rights at stake.

- Advances in genetic engineering and the cloning and freezing of gametes may soon offer an array of new social options and potential commercial opportunities. An arrangement that transforms human reproductive capacity into a commodity is, therefore, specially problematic at the present time.

- Public policy should discourage surrogate parenting. This goal should be achieved through legislation that declares the contracts void as against public policy. In addition, legislation should prohibit fees for surrogates and bar surrogate brokers from operating in New York State. Those measures are designed to eliminate commercial surrogacy and the growth of a business community or industry devoted to making money from human reproduction and the birth of children.

- The legislation proposed by the task force would not prohibit surrogate parenting arrangements when they are not commercial and remain undisputed. Existing law permits each step of the arrangement under these circumstances: a decision by a woman to be artificially inseminated or to have an embryo implanted, her voluntary decision after the child's birth to relinquish the child for adoption, and the child's adoption by the intended parents.

- Under existing adoption law, the intended parents would be permitted to pay reasonable expenses associated with pregnancy and childbirth to a mother who relinquishes her child for adoption. All such expenses must be approved by a court as part of an adoption proceeding.

- In custody disputes arising from surrogate parenting arrangements, the birth mother and her husband, if any, should be awarded custody unless the court finds, based on clear and convincing evidence, that the child's best interests would be served by an award of custody to the father and genetic mother. The court should award visitation and support obligations as it would under existing law in proceedings on these matters.

• To date, few programs have been conducted by the public or the private sector to prevent infertility. Programs to educate the public and health care professionals about the causes of infertility and the measures available for early detection and treatment could spare many couples from facing the problem. Both the government and the medical community should establish educational and other programs to prevent infertility. Resources should also be devoted to research about the causes and nature of infertility. •

Reprinted with the permission of the New York State Task Force on Life and the Law, 1988.

Excerpts from
Instruction on Respect for Human Life in Its Origin and on the Dignity of Procreation

Vatican Congregation or the Doctrine of the Faith, March, 1987

II. INTERVENTIONS UPON HUMAN PROCREATION

By *artificial procreation* or *artificial fertilization* are understood here the different technical procedures directed toward obtaining a human conception in a manner other than the sexual union of man and woman. This instruction deals with fertilization of an ovum in a test tube (in vitro fertilization) and artificial insemination through transfer into the woman's genital tracts of previously collected sperm.

A preliminary point for the moral evaluation of such technical procedures is constituted by the consideration of the circumstances and consequences which those procedures involve in relation to the respect due the human embryo. Development of the practice of *in vitro* fertilization has required innumerable fertilizations and destructions of human embryos. Even today, the usual practice presupposes a hyperovulation on the part of the woman: A number of ova are withdrawn, fertilized and then cultivated *in vitro* for some days. Usually not all are transferred into the genital tracts of the woman; some embryos, generally called "spare" parts, are destroyed or frozen. On occasion, some of the implanted embryos are sacrificed for various eugenic, economic or psychological reasons. Such deliberate destruction of human beings or their utilization for different purposes to the detriment of their integrity and life is contrary to the doctrine on procured abortion already recalled. The connection between in vitro fertilization and the voluntary destruction of human embryos occurs too often. This is significant: Through these procedures, with apparently contrary purposes, life and death are subjected to the decision of man, who thus sets himself up as the giver of life and death by decree. This dynamic of violence and domination may remain unnoticed by those very individuals who, in wishing to utilize this procedure, become subject to it themselves. The facts recorded and the cold logic which links them must be taken into consideration for a moral judgment on *in vitro* fertilization and embryo transfer: The abortion mentality which has made this procedure possible thus leads, whether one wants it or not, to man's domination over the life and death of his fellow human beings and can lead to a system of radical eugenics.

Nevertheless, such abuses do not exempt one from a further and thorough ethical study of the techniques of artificial procreation considered in themselves, abstracting as far as possible from the destruction of embryos produced *in vitro*.

The present instruction will therefore take into consideration in the first place the problems posed by heterologous artificial fertilization (II, 1–3),* and subsequently those linked with homologous artificial fertilization II, 4–6).**

Before formulating an ethical judgment on each of these procedures, the principles and values which determine the moral evaluation of each of them will be considered.

A. Heterologous Artificial Fertilization

1. Why must human procreation take place in marriage?

Every human being is always to be accepted as a gift and blessing of God. However, from the moral point of view a truly

---

*By the term heterogous artificial fertilization or procreation, the instruction means techniques used to obtain a human conception artificially by the use of gametes coming from at least one donor other than the spouses who are joined in marriage. Such techniques can be of two types:

a) *Heterologous "in vitro" fertilization and embryo transfer:* the technique used to obtain a human conception through the meeting in vitro of gametes taken from at least one donor other than the two spouses joined in marriage.

b) *Heterologous artificial insemination:* the technique used to obtain a human conception through the transfer into the genital tracts of the woman of the sperm previously collected from a donor other than her husband.

**By *artificial homologous fertilization or procreation*, the instruction means the technique used to obtain a human conception using the gametes of the two spouses joined in marriage. Homologous artificial fertilization can be carried out by two different methods:

a) *Homologous "in vitro" fertilization and embryo transfer:* the technique used to obtain a human conception through the meeting in vitro of the gametes of the spouses joined in marriage.

b) *Homologous artificial insemination:* the technique used to obtain a human conception through the transfer into the genital tracts of a married woman of the sperm previously collected from her husband.

responsible procreation vis-a-vis the unborn child must be the fruit of marriage.

For human procreation has specific characteristics by virtue of the personal dignity of the parents and of the children: The procreation of a new person, whereby the man and the woman collaborate with the power of the Creator, must be the fruit and the sign of the mutual self-giving of the spouses, of their love and of their fidelity. *The fidelity of the spouses in the unity of marriage involves reciprocal respect of their right to become a father and a mother only through each other.*

The child has the right to be conceived, carried in the womb, brought into the world and brought up within marriage: It is through the secure and recognized relationship to his own parents that the child can discover his own identity and achieve his own proper human development.

The parents find in their child a confirmation and completion of their reciprocal self-giving: The child is the living image of their love, the permanent sign of their conjugal union, the living and indissoluble concrete expression of their paternity and maternity.

By reason of the vocation and social responsibilities of the person, the good of the children and of the parents contributes to the good of civil society; the vitality and stability of society require that children come into the world within a family and that the family be firmly based on marriage.

The tradition of the church and anthropological reflection recognize in marriage and in its indissoluble unity the only setting worthy of truly responsible procreation.

2. Does heterologous artificial fertilization conform to the dignity of the couple and to the truth of marriage?

Through in vitro fertilization and embryo transfer and heterologous artificial insemination human conception is achieved through the fusion of gametes of at least one donor other than the spouses who are united in marriage. *Heterologous artificial fertilization is contrary to the unity of marriage, to the dignity of the spouses, to the vocation proper to parents, and to the child's right to be conceived and brought into the world in marriage and from marriage.*

Respect for the unity of marriage and for conjugal fidelity demands that the child be conceived in marriage; the bond existing between husband and wife accords the spouses, in an objective and inalienable manner, the exclusive right to become father and mother solely through each other. Recourse to the gametes of a third person in order to have sperm or ovum available constitutes a violation of the reciprocal commitment of the spouses and a grave lack in regard to that essential property of marriage which is its unity.

"**Human embryos obtained 'in vitro' are human beings and subjects with rights: Their dignity and right to life must be respected from the first moment of their existence. It is immoral to produce human embryos destined to be exploited as disposable biological material.**"

Heterologous artificial fertilization violates the rights of the child; it deprives him of his filial relationship with his parental origins and can hinder the maturing of his personal identity. Furthermore, it offends the common vocation of the spouses who are called to fatherhood and motherhood: It objectively deprives conjugal fruitfulness of its unity and integrity; it brings about and manifests a rupture between genetic parenthood, gestational parenthood and responsibility for upbringing. Such damage to the personal relationships within the family has repercussions on civil society: What threatens the unit and stability of the family is a source of dissension, disorder and injustice in the whole of social life.

These reasons lead to a negative moral judgment concerning heterologous artificial fertilization: Consequently, fertilization of a married woman with the sperm of a donor different from her husband and fertilization with the husband's sperm of an ovum not coming from his wife are morally illicit. Furthermore, the artificial fertilization of a woman who is unmarried or a widow, whoever the donor may be, cannot be morally justified.

The desire to have a child and the love between spouses who long to obviate a sterility which cannot be overcome in any other way constitute understandable motivations; but subjectively good intentions do not render heterologous artificial fertilization conformable to the objective and inalienable properties of marriage or respectful of the rights of the child and of the spouses.

3. Is "surrogate"* motherhood morally illicit?

No, for the same reasons which lead one to reject heterologous artificial fertilization:

For it is contrary to the unity of marriage and to the dignity of the procreation of the human person.

Surrogate motherhood represents an objective failure to meet the obligations of maternal love, of conjugal fidelity and of responsible motherhood; it offends the dignity and the right of the child to be conceived, carried in the womb, brought into the world and brought up by his own parents; it sets up, to the detriment of families, a division between the physical, psychological and moral elements which constitute those families.

5. Is homologous "in vitro" fertilization morally illicit?

The answer to this question is strictly dependent on the principles just mentioned. Certainly one cannot ignore the legitimate aspirations of sterile couples. For some, recourse to homologous "in vitro" fertilization and embryo transfer appears to be the only way of fullfilling their sincere desire for a child. The question is asked whether the totality of conjugal life in such situations is not sufficient to ensure the dignity proper to human procreation. It is acknowledged that in vitro fertilization and embryo transfer certainly cannot supply for the absence of sexual relation, and cannot be preferred to the specific acts of conjugal union, given the risks involved for the child and the difficulties of the procedure. But it is asked whether, when there is no otherway of overcoming the sterility which is a source of suffering, homologous in vitro fertilization may not constitute an aid, if not a form of therapy, whereby its moral licitness could be admitted.

The desire for a child—or at the very least an openness to the transmission of life—is a necessary prerequisite from the moral point of view for responsible human procreation. But this good intention is not sufficient for making a positive moral evaluation of *in vitro* fertilization between spouses. The process of *in vitro* fertilization and embryo transfer must be judged in itself and cannot borrow its definitive moral quality form the totality of conjugal life of which it becomes part nor from the conjugal acts which may precede or follow it.

It has already been recalled that in the circumstances in which it is regularly practiced *in vitro* fertilization and embryo transfer involves the destruction of human beings, which is something contrary to the doctrine on the illicitness of abortion previously mentioned. But even in a situation in which every precaution were taken to avoid the death of human embryos, homologous *in vitro* fertilization and embryo transfer dissociates from the conjugal act the actions which are directed to human fertilization. For this reason the very nature of homologous *in vitro* fertilization and embryo transfer also must be taken into account, even abstracting from the link with procured abortion.

Homologous *in vitro* fertilization and embryo transfer is brought about outside the bodies of the couple through actions of third parties whose competence and technical activity determine the success of the procedure. Such fertilization entrusts the life and identity of the embryo into the power of doctors and biologists and establishes the domination of technology over the origin and destiny of the human person. Such a relationship of domination is in itself contrary to the dignity and equality that must be common to parents and children.

Conception *in vitro* is the result of the technical action which presides over fertilization. *Such fertilization is neither in fact achieved nor positively willed as the expression and fruit of a specific act of the conjugal union. In homologous "in vitro" fertilization and embryo transfer, therefore, even if it is considered in the context of de facto existing sexual relations, the generation of the human person is objectively deprived of its proper perfection: namely, that of being the result and fruit of a conjugal act* in which the spouses can become "cooperators with God for giving life to a new person."

These reasons enable us to understand why the act of conjugal love is considered in the teaching of the church as the only setting worthy of human procreation. For the same reasons the so-called "simple case," i.e., a homologous *in vitro* fertilization and embryo transfer procedure that is free of any compromise with the abortive practice of destroying embryos and with masturbation, remains a technique which is morally illicit because it deprives human procreation of the dignity which is proper and connatural to it.

Certainly, homologous *in vitro* fertilization and embryo transfer fertilization is not marked by all that ethical negativity found in extraconjugal procreation; the family and marriage continue to constitute the setting for the birth and upbringing of the children. Nevertheless, in conformity with the traditional doctrine relating to the goods of marriage and the dignity of the person, *the church remains opposed from the moral point of view to homologous "in vitro" fertilization. Such fertilization is in itself illicit and in opposition to the dignity of procrea-*

tion and of the conjugal union, even when everything is done to avoid the death of the human embryo.

Although the manner in which human conception is achieved with in vitro fertilization and embryo transfer cannot be approved, every child which comes into the world must in any case be accepted as a living gift of the divine Goodness and must be brought up with love.

Index

Abortion
    19th Century views on, 31-32
    as a political issue, 32-34
    early medical opposition to, 33-34
    early criminal abortion laws, 34
    destruction of nontransferred fertilized eggs and, 50
    spontaneous abortion, 89-90
    See also In vitro fertilization; surrogacy.
Artificial insemination by donor (AID)
    viewed as adultery or fornication, 54-55
    lineage issues raised by, 55-56
    existing public policy on, 78
    success of, 98
    HIV virus and, 99
    cryopreservation and, 99-101
    risks associated with, 98-101.
Baby M
    background of, 183-87
    trial court decision, 187-90
    New Jersey Supreme Court decision, 190-94
    exploitation of women and, 195-96
    legal impacts of, 194-98
    See also Surrogacy.
Contraception
    in early 20th Century, 34-36
    recognized as a legitimate medical service, 37.
Contract motherhood. *See* Surrogacy.
Eugenics
    defined, 37
    history of, early 20th Century, 37-42
    political agendas and, 41
    biological determinism and, 41.
Experimentation with human subjects
    federal guidelines concerning, 155-56
    defined, 156-57.

Fetal status: 14, 20, 22-23, 26, 56-57, 72, 226-28.
Haldane, J.B.S.: 40.
Infertility
    defined, 92, 103-04, 112, 174
    causes of, 92-95, 112-13
    treatment success rates, 95, 97, 113, 116-17
    incidence of, 111-12
    risks associated with gonadotropin therapy, 97-98
    HIV virus and, 99-101
    Office of Technology Assessment report on, 111
    treatment costs associated with, 117
    public policy issues raised by, 119
    sexually transmitted diseases responsible for, 120-21
    prevention of, 120-21
    federal support for treatment of, 122-24
    informed consent standards for, 125-26
    legislation relating to, 126
    See also Artificial insemination by donor; Baby M; In vitro fertilization; Reproductive technologies; Surrogacy.
In vitro fertilization
    ethical principles applicable to, 72-76
    preembryos and spare embryos, 50, 69, 72-77 passim
    respect for persons principle applied to, 72-74, 166-68
    public policy issues, 70-72
    cryopreservation and, 74, 118
    beneficence principle applied to, 74-75
    justice principle applied to, 75-76
    selective killing in multiple pregnancy and, 76
    fetal abnormality rate in, 92
    gestational/social parenting distinction and, 104, 107, 248-49
    federal government policy and, 70-72, 122-23, 163-67, 246
    unnaturalness of, 50-51, 151-52, 228, 231
    informed consent and, 149-50, 152-53, 155, 166, 168
    research and therapeutic status of, 149, 152-55, 158-62, 164
    scarce medical resources and, 76, 161
    success rate of, 75, 158-61, 164-65, 168
    availability of, 165
    vulnerability of prospective parents and, 168
    See also Infertility.
Moral pluralism
    religion and, 45, 58
    approaches to public policy formulation and, 61-65 passim.

Index

National Commission for the Protection of Human Subjects: 67.
Pregnancy
    maternal experience associated with, 13, 20-24
    medical recognition of importance of maternal environment, 13-15, 19
    medical paternalism and, 15-16, 18
    as a relationship, 13-18, 18-26, 240-41.
Public policy
    relationship between law and morality in relation to, 58-59
    ethical criteria for the formulation of, 66-68.
Reproductive technologies
    patriarchal attitudes and, 9-10, 12
    gestational/social parenting distinction and, 12
    Catholic view on, 53-54
    Lutheran view on, 54
    Jehovah's Witness view on, 54-55
    religious concerns associated with, 46-47, 51-52, 58-63
    public policy issues, 59-60, 66-67
    public health insurance coverage for, 59, 60
    private health insurance coverage for, 59-61
    legislation regarding, 60-62
    alternative ethical perspectives on, 66
    laboratory procedures for, 68
    ethical issues raised by, 107-08
    psychosocial impacts of, 135-36, 140-43
    feminist response to, 133, 239-40.
Sexuality
    Victorian view toward, 131-32
    impact of reproductive technologies on, 133, 134-36, 137, 143-45.
Surrogacy
    ethical issues raised by, 77-78
    adoption and baby-selling laws impacted by, 78, 80
    British policy on, 78
    Baby M decision, 78
    public policy history, 78-79
    respect for persons principle applied to, 79-80
    informed consent dilemmas and, 79-80
    beneficence principle applied to, 80-83
    justice principle applied to, 83
    laboratory procedures, 104-05
    defined, 173, 175
    demographics, 175-76

economic costs of, 176
rationales of participants, 174-78 passim
public policy issues, 78-79, 178-80
unpredictable consequences of, 81-84
risks associated with, 198, 231
abortion and, 207-08, 209-11, 213-14
governmental regulation of, 179-80, 201
prenatal testing and, 203-07, 212
nontreatment of defective newborns and, 208-09, 214
Vatican policy on, 221, 225-26, 228
gestational/social parenting distinction and, 77, 81-83, 228-29, 232-33
anti-surrogacy views, 235-36
unnaturalness of, 229-34, 236
feminist response to, 237-38
social scientific research and, 243-44
family life and, 244-46
right to procreate and, 246-48
social consequences of, 253-54, 256-57
commercialization of human body and, 51, 57, 255-57
exploitation of women and, 51, 79, 83, 195-96, 237, 239, 240, 254-255, 257
child protection issues raised by, 81-83, 201-16 passim, 249-50, 257-58.

Made in the USA
San Bernardino, CA
04 March 2014